The Ultimate Math Survival Guide

- Part 1 -

Richard W. Fisher

IT IS ILLEGAL TO PHOTOCOPY THIS BOOK

The Ultimate Math Survival Guide, Part 1
ISBN 13: 978-0-9843629-5-0

Table of Contents

Whole Numbers and Integers

Fractions

Decimals and Percents

Answer Key

Notes to the Teacher or Parent

What sets this book apart from other books is its approach. It is not just a math book, but a system of teaching math. Each daily lesson contains three key parts: **Review Exercises**, **Helpful Hints**, and **Problem Solving**. Teachers have flexibility in introducing new topics, but the book provides them with the necessary structure and guidance. The teacher can rest assured that essential math skills in this book are being systematically learned.

This easy-to-follow program requires only fifteen or twenty minutes of instruction per day. Each lesson is concise and self-contained. The daily exercises help students to not only master math skills, but also maintain and reinforce those skills through consistent review - something that is missing in most math programs. Skills learned in this book apply to all areas of the curriculum, and consistent review is built into each daily lesson. Teachers and parents will also be pleased to note that the lessons are quite easy to correct.

This book is based on a system of teaching that was developed by a math instructor over a thirty-year period. This system has produced dramatic results for students. The program quickly motivates students and creates confidence and excitement that leads naturally to success.

Please read the following "How to Use This Book" section and let this program help you to produce dramatic results with your math students.

How to Use This Book

This book is best used on a daily basis. The first lesson should be carefully gone over with students to introduce them to the program and familiarize them with the format. It is hoped that the program will help your students to develop an enthusiasm and passion for math that will stay with them throughout their education.

As you go through these lessons every day, you will soon begin to see growth in the student's confidence, enthusiasm, and skill level. The students will maintain their mastery through the daily review.

Step 1

The students are to complete the review exercises, showing all their work. After completing the problems, it is important for the teacher or parent to go over this section with the students to ensure understanding.

Step 2

Next comes the new material. Use the "Helpful Hints" section to help introduce the new material. Be sure to point out that it is often helpful to come back to this section as the students work independently. This section often has examples that are very helpful to the students.

Step 3

It is highly important for the teacher to work through the two sample problems with the students before they begin to work independently. Working these problems together will ensure that the students understand the topic, and prevent a lot of unnecessary frustration. The two sample problems will get the students off to a good start and will instill confidence as the students begin to work independently.

Step 4

Each lesson has problem solving as the last section of the page. It is recommended that the teacher go through this section, discussing key words and phrases, and also key strategies. Problem solving is neglected in many math programs, and just a little work each day can produce dramatic results.

Step 5

Solutions are located in the back of the book. Teachers may correct the exercises if they wish, or have the students correct the work themselves.

Other Math Essentials Titles

Mastering Essential Math Skills: Book 1/Grades 4-5

Mastering Essential Math Skills: Book 2/Middle Grades/High School

Whole Numbers and Integers

Fractions

Decimals and Percents

Geometry

Problem Solving

Pre-Algebra Concepts

No-Nonsense Algebra

Try our free iphone app, Math Expert from Math Essentials

For more information go to www.mathessentials.net

The Secret to Success in Math

There are three clusters of math skills that every student needs to master. Students who learn and fully understand these essential topics can be considered algebra-ready. These skills are referred to as the Critical Foundations of Algebra.

Algebra-readiness is of huge importance. Algebra is the gateway subject to more advanced math, science, and technical classes. In turn, success in these classes will open a vast number of educational as well as career opportunities. In essence, success in math, and more specifically, Algebra, is a vital part of all students' education. Algebra-readiness will have a profound impact on success in school, college, career, and everyday life. Success in Algebra will open many doors for students. Unfortunately, those students who do not experience this success will find these same doors slammed shut.

Here are the Critical Foundations of Algebra:

- Whole Numbers—Students need to fully understand place value, and this must include a grasp of the meaning of the basic operations of addition, subtraction, multiplication, and division. They will also need the knowledge of how to apply the operations to problem solving. Instant recall of number facts is important. Whole number operations rest on the automatic recall of addition and related subtraction facts, and of multiplication and related division facts. These number facts are to math as the letters of the alphabet are to reading.

- Fractions—Students need to fully understand fractions, including decimals, and percents. This includes positive and negative fractions. They will need to be able to use all of these in problem solving. Fractions represent a major obstacle to a high percentage of students. Fractions, decimals, and percents need to be thoroughly understood.

- Some Aspects of Geometry and Measurements—Experience with similar triangles is directly relevant for the study of Algebra. Also, knowledge of slope of a line and linear functions is very important. Students should understand the properties of two and three–dimensional shapes and be able to determine perimeter area, volume and surface areas. They should also be able to find unknown lengths, angles, and areas. As with whole numbers and fractions, applying geometric skills to problem solving is essential.

For all these skills, conceptual understanding, computational fluency, and problem-solving skills are each essential.

The Critical Foundations of Algebra identified here are not meant to comprise a complete preschool-to-algebra curriculum. However, when these skills are mastered, success in Algebra will be assured.

The great news is that *The Ultimate Math Survival Guide: Part I* and *Part II* will ensure that students learn and master the Critical Foundations of Algebra!

The Ultimate Math Survival Guide Part I includes the following:

- Whole Numbers & Integers
- Fractions
- Decimals & Percents

The Ultimate Math Survival Guide Part II includes the following:

- Geometry
- Problem Solving
- Pre-Algebra

It should be noted that all of these skills are presented in a simple, easy-to-understand format. It is my belief, that ALL students can be successful in math! All that they need is the proper guidance.

Sincerely,
Richard W. Fisher
Author

Section 1

Whole Numbers and Integers

Review Exercises

Note to the students and teachers: This section will include review problems from all the topics covered in this book. Here are some simple problems with which to get started.

1.
$$\begin{array}{r} 314 \\ + 43 \\ \hline \end{array}$$

3.
$$\begin{array}{r} 7 \\ 2 \\ + 4 \\ \hline \end{array}$$

5. $7 + 4 + 5 =$

2.
$$\begin{array}{r} 603 \\ + 24 \\ \hline \end{array}$$

4.
$$\begin{array}{r} 426 \\ + 313 \\ \hline \end{array}$$

6.
$$\begin{array}{r} 22 \\ 16 \\ + 11 \\ \hline \end{array}$$

Helpful Hints

1. Line up numbers on the right side.
2. Add the ones first.
3. Remember to regroup when necessary.
4. Place commas in the answer when necessary.

* "Sum" means to total or add.

* Samples S1 and S2 may be worked together by teacher and students.

S1.
$$\begin{array}{r} 243 \\ 62 \\ + 514 \\ \hline \end{array}$$

S2.
$$\begin{array}{r} 345 \\ 423 \\ + 165 \\ \hline \end{array}$$

1.
$$\begin{array}{r} 42 \\ 56 \\ + 15 \\ \hline \end{array}$$

2.
$$\begin{array}{r} 516 \\ 54 \\ + 213 \\ \hline \end{array}$$

3.
$$\begin{array}{r} 517 \\ 326 \\ 143 \\ + 14 \\ \hline \end{array}$$

4.
$$\begin{array}{r} 813 \\ 236 \\ 17 \\ + 6 \\ \hline \end{array}$$

5.
$$\begin{array}{r} 543 \\ 227 \\ 312 \\ + 415 \\ \hline \end{array}$$

6.
$$\begin{array}{r} 5 \\ 23 \\ 416 \\ + 243 \\ \hline \end{array}$$

7. $415 + 436 + 317 =$

8. $23 + 34 + 16 + 47 =$

9. Find the sum of 443, 514, and 716.

10. Find the sum of 127, 23, 16, and 246.

1.
2.
3.
4.
5.
6.
7.
8.
9.
10.
Score

Problem Solving

There are 42 students in the fifth grade, 56 students in the sixth grade, and 36 students in the seventh grade. How many students are there altogether?

Review Exercises

1. $\begin{array}{r} 426 \\ +\ 23 \\ \hline \end{array}$

3. Find the sum of 23, 34, and 26.

5. $\begin{array}{r} 427 \\ 316 \\ 12 \\ +\ 423 \\ \hline \end{array}$

2. $\begin{array}{r} 526 \\ 42 \\ +\ 163 \\ \hline \end{array}$

4. $\begin{array}{r} 342 \\ 426 \\ +\ 325 \\ \hline \end{array}$

6. $\begin{array}{r} 36 \\ 324 \\ 573 \\ +\ 11 \\ \hline \end{array}$

| **Helpful Hints** | Use what you have learned to solve each problem.
REMEMBER: 1. Line up numbers on the right side.
2. Add the ones first.
3. Regroup when necessary
4. Place commas in the answer when necessary. | * "Sum" means add. |

* Samples S1 and S2 may be worked together by teacher and students.

S1. $\begin{array}{r} 426 \\ 343 \\ +\ 516 \\ \hline \end{array}$ S2. $\begin{array}{r} 16 \\ 247 \\ 346 \\ +\ 125 \\ \hline \end{array}$ 1. $\begin{array}{r} 702 \\ 337 \\ +\ 225 \\ \hline \end{array}$ 2. $\begin{array}{r} 23 \\ 45 \\ +\ 67 \\ \hline \end{array}$	1. 2. 3.

3. $\begin{array}{r} 627 \\ 423 \\ 504 \\ +\ 213 \\ \hline \end{array}$
4. $\begin{array}{r} 6 \\ 17 \\ 289 \\ +\ 363 \\ \hline \end{array}$
5. $\begin{array}{r} 724 \\ 427 \\ 395 \\ +\ 367 \\ \hline \end{array}$
6. $\begin{array}{r} 600 \\ 528 \\ 396 \\ +\ 27 \\ \hline \end{array}$

4.

5.

6.

7. $7 + 27 + 48 + 247 =$

7.

8. $763 + 29 + 372 + 16 =$

8.

9.

9. Find the sum of 96, 73, 44, and 75.

10.

10. Find the sum of 213, 426, 516, and 423.

Score

| **Problem Solving** | Monique earned $52 on Monday. On Tuesday, Wednesday, and Thursday she earned $56 each day. How much did she earn altogether? |

Review Exercises

1.   36
     47
     16
   + 92

2.   727
     423
   + 75

3.   824
     216
     724
   + 316

4.   967 + 843 + 96 =

5.   72 + 16 + 49 =

6.   Find the sum of 14, 15, 16, and 17.

Helpful Hints

When writing large numbers, place commas every three numerals, starting from the right. This makes them easier to read.

Example: 21 million, 234 thousand, 416

21,234,416

S1.   4,236
        147
     + 3,236

S2.   7,217
        685
     + 36,125

1.   4,736
      2,136
    + 3,523

2.   17,236
       3,175
     + 4,297

3.   1,312,516
    + 2,316,475

4.   7,137,236
        353,246
    +   7,506

5.   23,726
      17,532
       6,573
    + 23,247

6.   72,506
       1,343
      15,208
    + 5,123

7.   Find the sum of 1,342, 1,793, and 4,562.

8.   78,623 + 4,579 + 22,396 =

9.   72,214 + 16,738 + 29,143 + 17 =

10.   103,246 + 724,516 + 72,173 =

1.	
2.	
3.	
4.	
5.	
6.	
7.	
8.	
9.	
10.	
Score	

Problem Solving

One country has a population of 7,143,600. Another country has a population of 16,450,395. A third country has a population of 12,690,000. What is the total population of all three countries?

Review Exercises

1. 376
    493
    + 16

2. 23,463 + 7,764 + 21,976 =

3. 72,176
    43,267
    + 19,237

4. 27,967,204
    + 5,137,264

5. 95,623
    7,429
    19,234
    + 89,342

6. Find the sum of
    3,426,497 and
    7,136,095.

Helpful Hints

Use what you have learned to add the following problems. Practice reading your answers.

Example: 23 million, 106 thousand, 749

23,106,749

				1.

S1. 76,142
     7,913
     + 16,408

S2. 313,426
     2,462,417
     + 3,526,008

1. 27,167
    35,225
    + 16,342

2. 1,234,617
    3,426,124
    + 4,316,243

2. ___

3. ___

4. ___

3. 7,163
    2,476
    + 3,276

4. 17,226
    26,319
    15,314
    + 27,976

5.   16
      326
    7,764
    + 12,316

6. 76,174
    133,367
    243,776
    + 712,777

5. ___

6. ___

7. 76,724 + 55,726 + 5,723 + 74,126 =

7. ___

8. ___

8. 7,134,243 + 3,712,050 + 3,516,312 =

9. ___

9. Find the sum of 42,916, 47,993, and 6,716.

10. ___

10. 7 + 17 + 278 + 42,563 =

Score

Problem Solving

A school district has three high schools. One school has 2,463 students, another school has 3,985 students, and the third school has 1,596 students. What is the total number of high school students in the district.

15

Review Exercises

1.	33	2.	7,716	3.	7,124,563	4.	776
	45		12,727		5,236,907		397
	+ 76		+ 23,496		+ 7,369,704		+ 442

5. 753,476 + 569,736 +
7,842 + 39,673 =

6. Find the sum of 72, 96, 72, 9, and 88.

Helpful Hints	1. Line up the numbers on the right side. 2. Subtract the ones first. 3. Regroup when necessary. 4. It may be necessary to regroup more than once. 5. "Find the difference" and "how much more" means to subtract.	**Examples:**	$7\overset{8}{\cancel{9}}{}^1 3$ $-\ \ 75$ $\overline{718}$	$\overset{5\ \ 11}{\cancel{6}\cancel{2}3}$ $-\ 254$ $\overline{369}$

S1.	567	S2.	4,352	1.	339	2.	623
	- 183		- 2,171		- 24		- 52

3.	6,153	4.	5,231	5.	1,387	6.	7,383
	- 758		- 2,456		- 739		- 2,285

7. Find the difference between 763 and 47.

8. Subtract 527 from 3,916.

9. 7,249 - 779 =

10. 789 is how much more than 298?

1.
2.
3.
4.
5.
6.
7.
8.
9.
10.
Score

Problem Solving	846 students attend Vargas School and 943 students attend Hoover School. How many more students attend Hoover School than Vargas School?

Review Exercises

1. $\begin{array}{r} 726 \\ 849 \\ +\ 308 \\ \hline \end{array}$

3. $\begin{array}{r} 7,614 \\ -\ 1,557 \\ \hline \end{array}$

5. $\begin{array}{r} 5,613 \\ -\ \ 752 \\ \hline \end{array}$

6. $\begin{array}{r} 32,456,175 \\ 3,214,915 \\ +\ 6,318,425 \\ \hline \end{array}$

2. $\begin{array}{r} 721 \\ -\ 443 \\ \hline \end{array}$

4. Find the sum of 455, 673, and 998.

Helpful Hints

1. Use what you have learned to solve the following problems.
2. It may be necessary to regroup more than once.

"find the difference"
"is how much more"
"is how much less"

All Mean Subtraction

S1. $\begin{array}{r} 856 \\ -\ 297 \\ \hline \end{array}$

S2. $\begin{array}{r} 9,613 \\ -\ 2,247 \\ \hline \end{array}$

1. $\begin{array}{r} 715 \\ -\ 242 \\ \hline \end{array}$

2. $\begin{array}{r} 324 \\ -\ 149 \\ \hline \end{array}$

3. $\begin{array}{r} 3,137 \\ -\ 1,252 \\ \hline \end{array}$

4. $\begin{array}{r} 6,334 \\ -\ 2,175 \\ \hline \end{array}$

5. $\begin{array}{r} 7,123 \\ -\ 2,456 \\ \hline \end{array}$

6. $\begin{array}{r} 7,156 \\ -\ \ 877 \\ \hline \end{array}$

7. 27,346 - 15,472 =

8. Subtract 797 from 2,314.

9. 127 is how much less than 2,496?

10. Find the difference between 7,692 and 4,764.

1.	
2.	
3.	
4.	
5.	
6.	
7.	
8.	
9.	
10.	
Score	

Problem Solving

Ahmir earned 86 dollars on Friday. He earned 28 dollars less than this amount on Saturday. How much did he earn on Saturday?

Review Exercises

1. 395
    428
    + 376

3. 5,123 - 1,672 =

5. 376
    19
    426
    + 442

2. 7,162 - 396 =

4. Find the difference between 8,961 and 279.

6. 3,152 - 496 =

Helpful Hints	1. Line up the numbers on the right side. 2. Subtract the ones first. 3. It may be necessary to regroup more than once.	**Examples:** $$\begin{array}{r} {}^{6}\;{}^{10}\;{}^{9}\;{}_{1} \\ \cancel{7,103} \\ -\;\;\;677 \\ \hline 6,426 \end{array}$$ $$\begin{array}{r} {}^{5}\;{}^{9}\;{}^{9} \\ \cancel{6,000} \\ -\;1,634 \\ \hline 4,366 \end{array}$$

S1. 601
     - 356

S2. 700
     - 267

1. 90
    - 67

2. 705
    - 176

3. 4,012
    - 1,345

4. 700
    - 238

5. 8,000
    - 758

6. 5,207
    - 1,539

7. Find the difference between 13,012 and 10,796.

8. Subtract 7,689 from 9,026.

9. 70,023 - 63,345 =

10. What number is 5,021 less than 12,506?

1.
2.
3.
4.
5.
6.
7.
8.
9.
10.
Score

Problem Solving	A theatre has 1,150 seats. If 859 of them are taken, how many seats are empty?

Review Exercises

1.   500
    - 276

2.   3,196
       742
    + 276

3. Find the sum of 72, 95, 63, and 47.

4.   5,012
    - 763

5.   7,126
    - 3,147

6. 5,000 - 768 =

Helpful Hints	Use what you have learned to solve the following problems.	**Examples:**

$$\begin{array}{r} 4\ \ 11\ 9\ \ _1 \\ \cancel{5},\cancel{2}\cancel{0}1 \\ -\ \ 4\ 7\ 5 \\ \hline 4,7\ 2\ 6 \end{array} \qquad \begin{array}{r} 2\ \ 9\ \ 9 \\ \cancel{3},\cancel{0}\cancel{0}\cancel{0} \\ -\ 1,2\ 3\ 5 \\ \hline 1,7\ 6\ 5 \end{array}$$

S1.   6,002
     - 1,456

S2.   7,000
     - 967

1.   900
    - 77

2.   5,004
    - 2,576

3.   2,010
    - 1,346

4.   5,600
    - 797

5.   3,102
    - 1,453

6.   50,000
    - 7,542

7. 71,005 - 2,709 =

8. Subtract 2,511 from 7,005.

9. What number is 57 less than 500?

10. 76,500 - 29,308 =

1.
2.
3.
4.
5.
6.
7.
8.
9.
10.
Score

Problem Solving	A company earned 95,000 dollars this year. Last year the company earned 78,500 dollars. How much more did the company earn this year than last year?

Review Exercises

1. 375
    427
  + 28

2. 32,176
    3,724
  + 15,756

3. Find the sum of 24, 296, 752, and 897.

4. 7,121
  - 2,430

5. 7,001
  - 2,765

6. 4,000
  - 2,173

| **Helpful Hints** | Use what you have learned to solve the following problems. |

S1.   562
     7,124
    15,765
   + 1,456

S2. 4,001
   - 1,327

1. 516
     39
   + 77

2. 714
  - 253

3. 5,317
     967
   + 6,765

4. 7,000
   - 4,789

5. 6,102
   - 1,799

6.   29
     37
     46
   + 53

7. 23,076 - 13,097 =

8. Find the sum of 197, 368, and 427.

9. How much more is 921 than 143?

10. 72,176 + 7,996 + 72,199 =

| 1. |
| 2. |
| 3. |
| 4. |
| 5. |
| 6. |
| 7. |
| 8. |
| 9. |
| 10. |
| Score |

| **Problem Solving** | Pedro wants to buy a bike that costs $850. If he has saved $755, how much more does he need to be able to buy the bike? |

Review Exercises

1. 70
    + 26

2. 76,175 - 2,963 =

3. 325 + 72 + 765 + 89 =

6. 72,172
    5,347
    56
    + 2,347

4. 72,105
    - 7,367

5. Find the difference of
    7,723 and 10,197.

Helpful Hints	1. Use what you have learned to solve the following problems. 2. Line up numbers on the right. 3. Be careful when regrouping.

S1. 7,000
     - 1,345

S2. 37,673
     7,742
     + 7,396

1. 800
    - 124

2. 7,001
    - 1,237

3. 27
    347
    2,436
    + 7,998

4. 35,021
    - 7,130

5. 32 + 96 +
    33 + 37 =

6. 27
    99
    97
    68
    + 53

7. 976 + 279 + 342 + 196 =

8. 70,001 - 2,617 =

9. How much more is 7,526 than 3,017?

10. Find the sum of 22,463, 7,296, and 7,287.

1.
2.
3.
4.
5.
6.
7.
8.
9.
10.

Score

Problem Solving	71,503 people visited the museum on Monday. 80,106 people visited the museum on Friday. How many more people visited the museum on Friday than on Monday?

Review Exercises

1. $\begin{array}{r} 501 \\ -\ 267 \\ \hline \end{array}$

2. $\begin{array}{r} 334 \\ 79 \\ 617 \\ +\ 22 \\ \hline \end{array}$

3. $6{,}002 - 3{,}667 =$

6. $\begin{array}{r} 27{,}763 \\ 4{,}245 \\ 7{,}342 \\ +\quad 767 \\ \hline \end{array}$

4. $27 + 33 + 36 + 52 =$

5.   Find the difference of 8,012 and 796.

Helpful Hints	1. Line up numbers on the right. **Examples:** $\overset{1\ 1}{644} \atop \underline{\times\quad 3} \atop 1{,}932$ $\overset{\ \ 3\ 2}{6{,}0\,76} \atop \underline{\times\quad 4} \atop 24{,}304$
	2. Multiply the ones first.
	3. Regroup when necessary.
	4. "Product" means to multiply.

S1. $\begin{array}{r} 526 \\ \times\quad 3 \\ \hline \end{array}$

S2. $\begin{array}{r} 3{,}254 \\ \times\quad 6 \\ \hline \end{array}$

1. $\begin{array}{r} 67 \\ \times\quad 3 \\ \hline \end{array}$

2. $\begin{array}{r} 74 \\ \times\quad 6 \\ \hline \end{array}$

3. $\begin{array}{r} 427 \\ \times\quad 6 \\ \hline \end{array}$

4. $\begin{array}{r} 4{,}214 \\ \times\quad 6 \\ \hline \end{array}$

5. $\begin{array}{r} 3{,}056 \\ \times\quad 8 \\ \hline \end{array}$

6. $\begin{array}{r} 7{,}256 \\ \times\quad 6 \\ \hline \end{array}$

7. $7{,}036 \times 7 =$

8. $4 \times 3{,}849 =$

9.   Find the product of 8,762 and 6.

10.   Multiply 9 and 3,872.

1.
2.
3.
4.
5.
6.
7.
8.
9.
10.
Score

Problem Solving   If there are 365 days in each year, how many days are there in 7 years?

Review Exercises

1. 343
    x   4

2. 3,064
    x   7

3. 3,601 - 798 =

4. 6,000
    - 1,235

5. 375
    429
     63
    + 7

6. 763 x 4 =

Helpful Hints	Use what you have learned to solve the following problems.	*Remember Put commas in the answer if necessary and practice reading the answers.

S1. 7,234
     x   6

S2. 9,007
     x   8

1. 728
    x   6

2. 9,012
    x   9

3. 23,726
    x   7

4. 7,632
    x   6

5. 9,730
    x   9

6. 7,009
    x   4

7. Find the product of 8 and 2,643.

8. 8,012 x 7 =

9. 6 x 23,728 =

10. 12,429 x 6 =

1.	
2.	
3.	
4.	
5.	
6.	
7.	
8.	
9.	
10.	
Score	

Problem Solving	Ty needs a total of 450 points to win a price. If she has already earned 298 points, how many more points does she need to win the prize?

Review Exercises

1.
```
  725
x   6
```

2.
```
  3,012
x     9
```

3. 727 + 33 + 526 + 724 =

4. 23,102 is how much more than 7,256?

5.
```
  7,001
- 3,674
```

6.
```
    375
     47
    462
+   578
```

Helpful Hints

1. Line up numbers on the right.
2. Multiply the ones first.
3. Multiply the tens second.
4. Add the two products.
5. Place commas in the product if necessary.

Examples:
```
      43              437
x     32          x    26
      86             2622
+   1290          +  8740
   1,376            11,362
```

S1.
```
    43
x   24
```

S2.
```
   246
x   53
```

1.
```
    82
x   53
```

2.
```
    46
x   17
```

3.
```
    85
x   47
```

4.
```
   436
x   25
```

5.
```
   336
x   50
```

6.
```
   706
x   47
```

7. Find the product of 16 and 37.

8. 92 x 47 =

9. 763 x 45 =

10. 460 x 33 =

1.
2.
3.
4.
5.
6.
7.
8.
9.
10.
Score

Problem Solving

A school has 25 classrooms. If each classroom has 35 desks, how many desks are there altogether?

Review Exercises

1. 3,015
   - 176

2. 3,145
   x   7

3.   48
   x 36

4.   423
   x 25

5. 7,163
   5,427
   + 3,136

6. 7,562 - 1,999 =

Helpful Hints	Use what you have learned to solve the following problems. * Remember to put commas in your answer.

S1.   402
    x   26

S2.   356
    x   44

1.   26
   x 50

2.   56
   x 23

1.

2.

3.

3.   55
   x 46

4.   216
   x 64

5.   248
   x 76

6.   476
   x 75

4.

5.

6.

7. 92 x 103 =

7.

8. 468 x 26 =

8.

9.

9. Find the product of 65 and 70.

10.

10. Multiply 608 and 73.

Score

Problem Solving	3,216 students attend Washington Middle School. If 2,016 of the students are boys, how many girls attend the school?

Review Exercises

1.　42
　x 36

2.　407
　x 28

3.　400
　x 35

4.　7,612
　- 1,357

5.　36 + 37 + 19 + 62 =

6.　7,000 - 2,836 =

Helpful Hints

1. Line up numbers on the right.
2. Multiply the ones first.
3. Multiply the tens second.
4. Multiply the hundreds last.
5. Add the products.
6. Put commas in the product.

Examples:

```
    243              673
x   336          x   307
   1458             4711
   9720             0000
+ 72900          + 201900
  84,078          206,611
```

S1.　153
　x 423

S2.　724
　x 526

1.　247
　x 315

2.　246
　x 137

3.　244
　x 302

4.　364
　x 503

5.　543
　x 414

6.　269
　x 400

7.　521 x 337 =

8.　Find the product of 208 and 326.

9.　Multiply 500 and 822.

10.　444 x 366 =

1.	
2.	
3.	
4.	
5.	
6.	
7.	
8.	
9.	
10.	
Score	

Problem Solving

A factory can produce 3,050 cars per week.
How many cars can the factory produce in one year?
(Hint: There are 52 weeks in a year.)

Review Exercises

1. 7,632
      558
    3,627
  +   36

3. 2,172
  x   6

5.   263
  x   30

2. 3,000
  - 2,176

4.   46
  x 23

6.   526
  x 704

Helpful Hints	Use what you have learned to solve the following problems. 1. Put commas in the answers. 2. Practice reading the answers.

S1.   444
   x 225

S2.   350
   x 906

1.   906
  x 717

2.   500
  x 743

3.   323
  x 435

4.   648
  x 755

5.   672
  x 986

6.   246
  x 932

7. 219 x 406 =

8. 700 x 610 =

9. 763 x 908 =

10. Multiply 723 and 847.

1.	
2.	
3.	
4.	
5.	
6.	
7.	
8.	
9.	
10.	
Score	

Problem Solving	Each bus can hold 112 students. How many students can 7 buses hold?

Review Exercises

1. 408
    x   6

2. Find the product
    of 24 and 36.

3. 7,665
    8,327
    + 9,342

4. 3,121
    - 2,344

5. 742
    x   6

6. 408
    x   27

| **Helpful Hints** | Use what you have learned to solve the following problems. |

S1. 426
     x   25

S2. 613
     x   425

1. 37
    x   5

2. 705
    x   6

3. 2,346
    x   7

4. 58
    x   72

5. 245
    x   63

6. 128
    x   547

7. Find the product of 15 and 726.

8. 700 x 657 =

9. 33 x 610 =

10. 308 x 407 =

1.
2.
3.
4.
5.
6.
7.
8.
9.
10.
Score

| **Problem Solving** | If there are 1,440 minutes in a day, how many minutes are there in a week? |

Review Exercises

1. 365 + 19 + 342 =

2. 627
   x   5

3. 7,136
   -   807

4. 214
   x   7

5. 206
   x 77

6. 364
   x 500

Helpful Hints	1. Use what you have learned to solve the following problems. 2. Carefully line up the numbers. 3. Put commas in answers when necessary 4. Practice reading the answers.

S1. 6,007
    x   5

S2. 347
    x 55

1. 404
   x   5

2. 7,002
   x   6

1.
2.
3.
4.
5.
6.
7.
8.
9.
10.
Score

3. 8,916
   x   7

4.   96
   x 80

5. 718
   x 63

6. 424
   x 726

7. 3 x 720 =

8. 9 x 7,856 =

9. 67 x 715 =

10. 208 x 763 =

Problem Solving	726 people attend the theatre on Saturday. On Sunday 827 people attended the theatre. How many more people attended the theatre on Sunday than on Saturday?

Review Exercises

1. 624
   $-$ 317

2. 347
   x   5

3. 72,196
   $+$ 16,429

4. Find the difference between 7,912 and 995.

5. 136
   x   22

6. 200 x 309 =

Helpful Hints

1. Divide
2. Multiply
3. Subtract
4. Begin Again

Examples:

$$\begin{array}{r} 15\,r2 \\ 3\overline{)47} \\ -3\!\downarrow \\ \hline 17 \\ -15 \\ \hline 2 \end{array}$$

$$\begin{array}{r} 9\,r5 \\ 6\overline{)59} \\ -54 \\ \hline 5 \end{array}$$

Remember! The remainder must be less than the divisor.

S1. $4\overline{)17}$

S2. $5\overline{)49}$

1. $3\overline{)74}$

2. $8\overline{)47}$

3. $6\overline{)78}$

4. $5\overline{)93}$

5. $7\overline{)84}$

6. $5\overline{)79}$

7. 77 ÷ 8 =

8. 96 ÷ 4 =

9. $\dfrac{65}{4}$

10. $\dfrac{47}{2}$

1.	
2.	
3.	
4.	
5.	
6.	
7.	
8.	
9.	
10.	
Score	

Problem Solving

Pencils come in boxes of 72. If the pencils are divided equally among 4 students, how many pencils will each student receive?

Review Exercises

1. $2\overline{)17}$ 2. $5\overline{)89}$ 3. $7\overline{)45}$

4. $205 - 99 =$ 5. $\begin{array}{r} 165 \\ \times\ \ 7 \\ \hline \end{array}$ 6. $\begin{array}{r} 47 \\ \times\ 33 \\ \hline \end{array}$

Helpful Hints	Use what you have learned to solve the following problems. 1. Divide 2. Multiply 3. Subtract 4. Begin Again 5. REMEMBER! The remainder must be less than the divisor.

S1. $4\overline{)73}$ S2. $6\overline{)27}$ 1. $3\overline{)29}$ 2. $3\overline{)45}$

1.
2.
3.

3. $8\overline{)93}$ 4. $2\overline{)97}$ 5. $7\overline{)60}$ 6. $5\overline{)29}$

4.
5.
6.

7. $77 \div 5 =$ 8. $\dfrac{99}{4}$

7.
8.

9. $33 \div 4 =$ 10. $65 \div 5 =$

9.
10.
Score

Problem Solving	Mr. Johnson's salary is $6,500 per month. How much does he earn per year? (Hint: How many months are in a year?)

Review Exercises

1. $6\overline{)38}$

2. $5\overline{)69}$

3. $7{,}103 - 2{,}617 =$

4. $526 \times 7 =$

5. $\begin{array}{r} 164 \\ \times\ \ 23 \\ \hline \end{array}$

6. $73 + 16 + 57 + 76 =$

Helpful Hints

1. Divide
2. Multiply
3. Subtract
4. Begin Again

Examples:

$$
\begin{array}{r}
171\ r2 \\
3\overline{)515} \\
-3 \\
\hline
21 \\
-21 \\
\hline
05 \\
-3 \\
\hline
2
\end{array}
\qquad
\begin{array}{r}
203 \\
4\overline{)812} \\
-8 \\
\hline
01 \\
-0 \\
\hline
12 \\
-12 \\
\hline
0
\end{array}
$$

REMEMBER!
The remainder must be less than the divisor.

S1. $2\overline{)523}$

S2. $6\overline{)274}$

1. $3\overline{)972}$

2. $2\overline{)419}$

3. $7\overline{)933}$

4. $6\overline{)727}$

5. $8\overline{)916}$

6. $6\overline{)850}$

7. $3\overline{)936}$

8. $5\overline{)607}$

9. $3\overline{)776}$

10. $8\overline{)717}$

1.

2.

3.

4.

5.

6.

7.

8.

9.

10.

Problem Solving

A bakery produced 324 cookies. If 9 cookies are placed into each package, how many packages will the bakery need?

Score

Review Exercises

1. $3\overline{)57}$

2. $7\overline{)89}$

3. $5\overline{)616}$

4. $5\overline{)374}$

5. $7\overline{)924}$

6. $7\overline{)450}$

Helpful Hints	Use what you have learned to solve the following problems.

S1. $2\overline{)375}$ S2. $4\overline{)377}$ 1. $8\overline{)916}$ 2. $8\overline{)630}$

3. $7\overline{)679}$ 4. $8\overline{)976}$ 5. $4\overline{)508}$ 6. $6\overline{)248}$

7. $3\overline{)400}$ 8. $7\overline{)693}$ 9. $5\overline{)778}$ 10. $6\overline{)609}$

1.
2.
3.
4.
5.
6.
7.
8.
9.
10.
Score

Problem Solving	The Martinez family took a vacation and drove 450 miles each day for 12 days. How many miles did they drive during their vacation?

Review Exercises

1. 3⟌69

3. 213
    x 47

5. 900 x 614 =

2. 5⟌667

4.   7,710
    x 697

6.   712
      14
    + 47

Helpful Hints

1. Divide
2. Multiply
3. Subtract
4. Begin Again

Examples:

```
      1708
  3⟌5124
   - 3↓
     21
   - 21↓
     02
    - 0↓
     24
   - 24
      0
```

```
       448 r2
  4⟌1794
   - 16↓
     19
   - 16↓
     34
   - 32
      2
```

REMEMBER!
The remainder
must be less than
the divisor.

S1. 3⟌9052   S2. 5⟌3352   1. 2⟌9235   2. 2⟌7362

3. 6⟌6821   4. 5⟌2240   5. 5⟌7666   6. 4⟌7123

7. 5⟌61,235   8. 2⟌24,362   9. 4⟌41,236   10. 5⟌22,014

1.
2.
3.
4.
5.
6.
7.
8.
9.
10.

Problem Solving

5 students opened a business and earned $16,550.
If they divided the money equally, how much would
each student receive?

Score

Review Exercises

1. $3\overline{)526}$

2. $3\overline{)605}$

3. $7\overline{)2634}$

4. $5\overline{)6175}$

5. $6\overline{)13,252}$

6. $7,100 - 768 =$

Helpful Hints	1. Use what you have learned to solve the following problems. 2. Remainders must be less than the divisor. 3. Zeroes may sometimes appear in the quotient.

S1. $6\overline{)3976}$ S2. $5\overline{)13,976}$ 1. $3\overline{)4123}$ 2. $5\overline{)1726}$

3. $8\overline{)8902}$ 4. $6\overline{)5324}$ 5. $6\overline{)9309}$ 6. $8\overline{)8016}$

7. $3\overline{)16,723}$ 8. $7\overline{)22,305}$ 9. $4\overline{)61,053}$ 10. $8\overline{)37,716}$

1.

2.

3.

4.

5.

6.

7.

8.

9.

10.

Score

Problem Solving	Yana took a 3-day hiking trip. The first day she hiked 13 miles, the second day she hiked 14 miles, and the third day she hiked 17 miles. How many miles did she hike altogether during the trip?

Review Exercises

1. 337
    96
    + 349

2. 32,105
    - 6,737

3. 427
    x   26

4. 4⟌500

5. 7⟌1356

6. Find the sum of 39,764 and 79,743.

| **Helpful Hints** | 1. Divide 2. Multiply 3. Subtract 4. Begin again | * Remainders must be less than the divisor. * Zeroes may sometimes appear in the quotient. |

S1. 3⟌901   S2. 7⟌7563   1. 5⟌402   2. 5⟌8500

3. 4⟌8413   4. 8⟌3566   5. 6⟌1526   6. 5⟌6000

7. 4⟌8009   8. 9⟌8765   9. 7⟌6329   10. 4⟌3217

1.
2.
3.
4.
5.
6.
7.
8.
9.
10.

Score

Problem Solving

A ream of paper contains 500 sheets.
How many sheets of paper are there in 25 reams?

Review Exercises

1. Find the difference between 7,026 and 4,567.

2.
$$637$$
$$\times\ 502$$

3. $7\overline{)600}$

4. $5\overline{)5003}$

5. $2\overline{)3051}$

6. $7\overline{)1969}$

Helpful Hints

Use what you have learned to solve the following problems.

S1. $3\overline{)369}$

S2. $6\overline{)8772}$

1. $5\overline{)307}$

2. $9\overline{)157}$

3. $6\overline{)1213}$

4. $7\overline{)7012}$

5. $6\overline{)9736}$

6. $4\overline{)1398}$

7. $8\overline{)13,423}$

8. $5\overline{)14,387}$

9. $6\overline{)71,234}$

10. $5\overline{)15,355}$

1.

2.

3.

4.

5.

6.

7.

8.

9.

10.

Score

Problem Solving

A dairy produced 4,448 gallons of milk. If the milk is put into containers that hold 8 gallons each, how many containers will be needed?

Review Exercises

1. $6\overline{)718}$

2. $2,007 - 865 =$

3.
$$\begin{array}{r} 324 \\ \times\ 700 \\ \hline \end{array}$$

4. $5\overline{)5007}$

5. $6\overline{)1399}$

6.
$$\begin{array}{r} 3,966 \\ 7,723 \\ +\ 4,564 \\ \hline \end{array}$$

Helpful Hints	1. Divide 2. Multiply 3. Subtract 4. Begin again	Examples:

$$\begin{array}{r} 12\ \text{r}49 \\ 60\overline{)769} \\ -\ 60\downarrow \\ \hline 169 \\ -\ 120 \\ \hline 49 \end{array} \qquad \begin{array}{r} 44\ \text{r}5 \\ 40\overline{)1765} \\ -\ 160\downarrow \\ \hline 165 \\ -\ 160 \\ \hline 5 \end{array}$$

S1. $30\overline{)176}$

S2. $40\overline{)5732}$

1. $70\overline{)238}$

2. $50\overline{)376}$

3. $50\overline{)438}$

4. $70\overline{)829}$

5. $30\overline{)1682}$

6. $20\overline{)1396}$

7. $40\overline{)8972}$

8. $90\overline{)9096}$

9. $40\overline{)3786}$

10. $40\overline{)2396}$

1.

2.

3.

4.

5.

6.

7.

8.

9.

10.

Score

Problem Solving

A store puts eggs into boxes of a dozen. If there are 3,072 eggs, how many boxes will be needed?

Review Exercises

1. $2\overline{)69}$

2. $7\overline{)7019}$

3. $367 + 246 +$ $721 + 243 =$

4. $36 \times 304 =$

5. $6,108 - 976 =$

6. $3\overline{)6123}$

Helpful Hints	1. Use what you have learned to solve the following problems. 2. Zeroes may sometimes appear in the quotient. 3. Remember that remainders must be less than the divisor.

S1. $50\overline{)137}$ S2. $60\overline{)7612}$ 1. $20\overline{)396}$ 2. $20\overline{)187}$

3. $30\overline{)2312}$ 4. $30\overline{)7396}$ 5. $70\overline{)5724}$ 6. $70\overline{)9988}$

7. $50\overline{)9976}$ 8. $40\overline{)1766}$ 9. $80\overline{)2209}$ 10. $50\overline{)12,076}$

1.	
2.	
3.	
4.	
5.	
6.	
7.	
8.	
9.	
10.	
Score	

Problem Solving	A school has 24 classes. Each class contains 32 students. How many students are there altogether in the school?

Review Exercises

1. 463
    87
    + 496

2. 7,015
    - 1,247

3. 963 - 713 =

4. 60 ⟌ 729

5. 70 ⟌ 367

6. 40 ⟌ 5276

| **Helpful Hints** | Sometimes it is easier to mentally round the divisor to the nearest power of ten. | Examples: $\begin{array}{r} 32\ r11 \\ 22\overline{\smash{)}715} \\ -\ 66\!\downarrow \\ \hline 55 \\ -\ 44 \\ \hline 11 \end{array}$ | Think of: 20 ⟌ 715 |

S1. 32 ⟌ 673   S2. 32 ⟌ 279   1. 41 ⟌ 936   2. 28 ⟌ 617

3. 58 ⟌ 697   4. 83 ⟌ 762   5. 34 ⟌ 826   6. 31 ⟌ 976

7. 62 ⟌ 759   8. 53 ⟌ 609   9. 21 ⟌ 936   10. 42 ⟌ 866

1.

2.

3.

4.

5.

6.

7.

8.

9.

10.

Score

| **Problem Solving** | A car traveled 448 miles. For each 32 miles it traveled, it consumed one gallon of gas. How many gallons of gas did it consume in traveling 448 miles? |

Review Exercises

1.   326
   x 35

2.   324
   x 500

3.   27 + 89 + 76 + 14 =

5.   50|826

6.   90|7372

4.   3,000 - 756 =

Helpful Hints	Use what you have learned to solve the following problems. *Remember: Sometimes mentally rounding the divisor helps to make the problem easier.	Examples: $\begin{array}{r} 36 \\ 27\overline{)998} \\ -81\downarrow \\ \hline 188 \\ -162 \\ \hline 26 \end{array}$ Think of: 30|998

S1. 41|912   S2. 51|367   1. 52|706   2. 39|876

3. 47|613   4. 22|796   5. 12|393   6. 25|618

7. 49|536   8. 22|684   9. 28|976   10. 42|750

1.

2.

3.

4.

5.

6.

7.

8.

9.

10.

Score

Problem Solving	A theater had 12 rows of seats with 16 seats in a row. If only 7 seats were empty, how many seats were occupied?

Review Exercises

1. $30\overline{)98}$ 2. $60\overline{)793}$ 3. $60\overline{)483}$

4. $32\overline{)426}$ 5. $32\overline{)275}$ 6. $28\overline{)405}$

| **Helpful Hints** | Sometimes it is necessary to correct your estimate. | Examples: $\dfrac{6}{63\overline{)374}}$ $\underline{-378}$ - too large | $63\overline{)374}^{\,5r60}$ $\underline{-315}$ 60 |

S1. $74\overline{)293}$ S2. $43\overline{)821}$ 1. $38\overline{)197}$ 2. $88\overline{)522}$

3. $18\overline{)987}$ 4. $22\overline{)178}$ 5. $32\overline{)163}$ 6. $14\overline{)886}$

7. $34\overline{)649}$ 8. $42\overline{)829}$ 9. $36\overline{)721}$ 10. $18\overline{)778}$

1.

2.

3.

4.

5.

6.

7.

8.

9.

10.

Score

| **Problem Solving** | A machine can produce 75 parts in one hour. How many parts can it produce in 12 hours? |

Review Exercises

1. 26
    x 3

2. 2,476
    x 7

3. 64
    x 23

4. 207
    x 25

5. 627
    x 400

6. 763
    x 509

Helpful Hints	Use what you have learned to solve the following problems. *Remember: Sometimes it is necessary to correct your estimate.

S1. 29$\overline{)862}$ S2. 22$\overline{)830}$ 1. 27$\overline{)851}$ 2. 25$\overline{)805}$

3. 19$\overline{)572}$ 4. 23$\overline{)476}$ 5. 31$\overline{)927}$ 6. 12$\overline{)588}$

7. 18$\overline{)787}$ 8. 42$\overline{)829}$ 9. 14$\overline{)886}$ 10. 21$\overline{)178}$

1.

2.

3.

4.

5.

6.

7.

8.

9.

10.

Score

Problem Solving	Children's tickets to a show are 5 dollars each and adult tickets are 7 dollars each. What would the total price be for 5 children's tickets and 4 adult tickets?

Review Exercises

1.
```
   721
 - 356
```

2.   755 - 288 =

3.
```
   60,123
 - 29,465
```

4.
```
   779
   643
   247
 +  96
```

5.   637 + 975 + 734 + 623 =

6.   7,256 is how much more than 5,909?

Helpful Hints

Use what you have learned to solve the following problems.

*Remember: Mentally round your divisor to the nearest power of ten.

Example:
```
        216 r20
   32 ) 6932
      - 64↓
         53
       - 32↓
         212
       - 192
          20
```

Think of:
```
   30 ) 6932
```

S1.   43) 9250   S2.   32) 6635   1.   21) 2645   2.   51) 7563

3.   42) 9005   4.   31) 5126   5.   27) 8056   6.   25) 7651

7.   81) 2654   8.   22) 1765   9.   28) 3976   10.   12) 9006

1.

2.

3.

4.

5.

6.

7.

8.

9.

10.

Problem Solving

A car traveled at the speed of 55 miles per hour for 8 hours. What was the total distance traveled?

Score

44

Review Exercises

1. $3\overline{)106}$ 2. $7\overline{)6105}$ 3. $90\overline{)9672}$

4. $90\overline{)3807}$ 5. $42\overline{)506}$ 6. $42\overline{)369}$

| **Helpful Hints** | Use what you have learned to solve the following problems. | | |

S1. $34\overline{)1637}$ S2. $32\overline{)6795}$ 1. $41\overline{)3876}$ 2. $51\overline{)3962}$

3. $63\overline{)7196}$ 4. $81\overline{)7642}$ 5. $78\overline{)9053}$ 6. $58\overline{)3012}$

7. $25\overline{)7942}$ 8. $49\overline{)8006}$ 9. $12\overline{)4813}$ 10. $72\overline{)3099}$

1.

2.

3.

4.

5.

6.

7.

8.

9.

10.

Score

Problem Solving

A farmer has 423 eggs. If he wants to put them into cartons that hold 36 eggs, how many full cartons will he have? How many eggs will be left over?

Review Exercises

1.　　72,096
　　　　7,423
　　+　27,564

2.　　7,505
　　-　3,176

3.　Find the sum of
　　79,216 and 76,719.

4.　Find the product of
　　725 and 301.

5.　Find the difference of
　　6,000 and 890.

6.　　336
　　x　　22

Helpful Hints	Use what you have learned to solve the following problems.	*Mentally round 2-digit divisors when helpful. *Remainders must be less than the divisor.

				1.
S1.　$81\overline{)7716}$	S2.　$28\overline{)1561}$	1.　$2\overline{)77}$	2.　$6\overline{)1989}$	2.
				3.
				4.
3.　$5\overline{)1397}$	4.　$40\overline{)647}$	5.　$90\overline{)706}$	6.　$90\overline{)3762}$	5.
				6.
				7.
7.　$38\overline{)966}$	8.　$23\overline{)278}$	9.　$61\overline{)3344}$	10.　$32\overline{)8096}$	8.
				9.
				10.
				Score

Problem Solving	There were 96 students on a bus. At one stop 28 students got off. At the next stop 17 students got off. How many students were left on the bus?

Review Exercises

1. $2\overline{)68}$ 2. $3\overline{)1964}$ 3. $30\overline{)279}$

4. $30\overline{)1367}$ 5. $62\overline{)707}$ 6. $62\overline{)4537}$

Helpful Hints	Use what you have learned to solve the following problems.	

				1.
S1. $61\overline{)5872}$	S2. $27\overline{)2096}$	1. $5\overline{)97}$	2. $8\overline{)2605}$	2.
				3.
				4.
3. $9\overline{)7002}$	4. $60\overline{)967}$	5. $60\overline{)396}$	6. $80\overline{)6776}$	5.
				6.
				7.
				8.
7. $76\overline{)396}$	8. $76\overline{)962}$	9. $76\overline{)1796}$	10. $76\overline{)9908}$	9.
				10.
				Score

Problem Solving   Susan works 40 hours per week. If she is paid 12 dollars per hour, how much does she earn in 2 weeks?

       47

Review of All Whole Number Operations

1.
```
   456
    27
+  626
```

2.
```
   819
   746
   696
+  638
```

3. $7,964 + 895 + 72,528 =$

4. $7,096 + 2,716 + 779 + 84 =$

5. $7,009 + 868 + 19 + 578 =$

6.
```
   742
-  397
```

7.
```
  6,291
- 3,476
```

8. $9,051 - 2,766 =$

9. $8,000 - 3,988 =$

10. $9,051 - 2,766 =$

11.
```
   87
x   3
```

12.
```
  6,542
x     7
```

13.
```
   49
x  63
```

14.
```
   809
x   76
```

15.
```
   409
x  278
```

16. $3\overline{)425}$

17. $6\overline{)1697}$

18. $40\overline{)867}$

19. $53\overline{)7976}$

20. $38\overline{)1698}$

1.
2.
3.
4.
5.
6.
7.
8.
9.
10.
11.
12.
13.
14.
15.
16.
17.
18.
19.
20.

Final Review of All Whole Number Operations

1. 767
    455
    396
    + 228

2. 72,367
    9,768
    + 7,709

3. 16,221 + 6,872 + 9,796 + 15 =

4. 7 + 77 + 777 + 7,777 =

5. 9,697 + 7,636 + 8,964 =

6. 700
    - 265

7. 9,001
    - 786

8. 52,153 - 7,654 =

9. 5,000 - 2,357 =

10. 72,591 - 16,784 =

11. 308
     x   7

12. 7,010
     x   9

13. 87
     x 29

14. 978
     x 39

15. 963
     x 801

16. 7$\overline{)412}$

17. 6$\overline{)1392}$

18. 40$\overline{)239}$

19. 63$\overline{)8974}$

20. 29$\overline{)17,236}$

1.
2.
3.
4.
5.
6.
7.
8.
9.
10.
11.
12.
13.
14.
15.
16.
17.
18.
19.
20.

Review Exercises

1. 775 + 639 + 426 =

2. 7,163
    -   269

3. Find the difference between
    7,455 and 3,672.

4. 7⟌1637

5. 70⟌367

6. 40⟌719

Helpful Hints

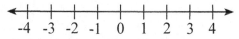

-4 -3 -2 -1 0 1 2 3 4

Integers to the left of zero are negative and less than zero. Integers to the right of zero are positive and greater than zero. When two integers are on a number line, the one farthest to the right is greater.

Hint: Always find the sign of the answer first when working these problems.

Examples: The sum of two negatives is a negative.

-7 + -5 = -   7
                 + 5
(The sign is negative.)   12 = (-12)

When adding a negative and a positive, the sign is the same as the integer farthest from zero, then subtract.

-7 + 9 = +   9
                  - 7
(The sign is positive.)   2 = (+2)

S1. -7 + 13 =

S2. -16 + -7 =

1. -17 + 26 =

2. -14 + -4 =

3. 56 + -72 =

4. -14 + -13 =

5. -9 + 16 =

6. -47 + 85 =

7. -102 + -78 =

8. 65 + -94 =

9. -23 + -47 =

10. 65 + -85 =

1.

2.

3.

4.

5.

6.

7.

8.

9.

10.

Problem Solving

There are 320 students going on a field trip. If each bus holds 55 students, how many buses will be needed for the field trip?

Score

Review Exercises

1. -32 + 18 =

2. 17 + -12 =

3. -32 + -46 =

4. 453
    x 327

5. 32|409

6. 32|197

Helpful Hints	Use what you have learned to solve the following problems. *Remember to find the sign of the answer first.

S1. -96 + 105 =

S2. -99 + -86 =

1. -67 + 58 =

2. -235 + -701 =

3. -56 + 19 =

4. -95 + -46 =

5. -163 + 200 =

6. -423 + 208 =

7. -525 + -376 =

8. 924 + -1,023 =

9. -346 + -295 =

10. 650 + -496 =

1.

2.

3.

4.

5.

6.

7.

8.

9.

10.

Score

Problem Solving	Maria bought a CD that cost 19 dollars. If she paid with a 50 dollar bill, how much change will she receive?

Review Exercises

1.   3,009
       x   7

2.   767
         96
      +   394

3.   72,052
      -   13,654

4. -75 + 60 =

5. -92 + -96 =

6. -105 + 142 =

Helpful Hints	When adding more than two integers, group the negatives and positives separately, then add.	**Examples:** $-6 + 4 + -5 =$ $-11 + 4 = -$ (Sign is negative.) $\begin{array}{r} 11 \\ -\ 4 \\ \hline 7 \end{array} = \boxed{-7}$	$7 + -3 + -8 + 6 =$ $-11 + 13 = +$ (Sign is positive.) $\begin{array}{r} 13 \\ -\ 11 \\ \hline 2 \end{array} = \boxed{+2}$

S1.   -4 + 7 + -5 =

S2.   -7 + 6 + -8 + 4 =

1.   -7 + -5 + 15 =

2.   9 + -4 + -8 =

3.   -16 + 22 + -12 =

4.   8 + -6 + 4 + 7 =

5.   -22 + 40 + -24 =

6.   -12 + 8 + -10 + 6 =

7.   -12 + -7 + -14 =

8.   -7 + 8 + -3 + -7 =

9.   -40 + 18 + -16 + 12 =

10.   -84 + 30 + -35 =

1.

2.

3.

4.

5.

6.

7.

8.

9.

10.

Score

Problem Solving

A student received test scores of 84, 90, and 93. What was his average score?

Review Exercises

1. $40\overline{)967}$

2. $40\overline{)384}$

3. $32\overline{)697}$

4. $32\overline{)1987}$

5. $-4 + 8 + -9 + 7 =$

6. $-35 + 40 + -77 =$

Helpful Hints	Use what you have learned to solve the following problems. Remember to group the negatives and positives separately, then add.

S1. $-7 + 9 + -8 =$

S2. $-5 + 12 + -13 + 15 =$

1. $-6 + -8 + 7 =$

2. $15 + -7 + -30 =$

3. $-50 + 17 + -22 =$

4. $16 + -12 + 18 + -8 =$

5. $-52 + 32 + -14 =$

6. $-20 + 14 + -12 + 32 =$

7. $62 + -101 + 15 =$

8. $-35 + -36 + -37 =$

9. $-12 + 21 + -16 + 40 =$

10. $-92 + 58 + -23 =$

1.

2.

3.

4.

5.

6.

7.

8.

9.

10.

Score

Problem Solving	A business needs 600 postcards to mail to customers. If postcards come in packages of 25, how many packages does the business need to buy?

Review Exercises

1. -17 + 5 + -16 =

2. -5 + -6 + 7 + 12 =

3. -14 + 25 + -32 + 6 =

4. 36 + 72 + 14 + 96 + 23 =

5. 2,015 - 786 =

6.
```
   786
x   22
```

Helpful Hints

* To subtract integers means to add to its opposite.   **Examples:**

```
-3 - -8 =           8
-3 + 8 = +        - 3
(Sign is positive.)  5 = (+5)
```

```
8 - 10 =           10
8 + -10 = -        - 8
(Sign is negative.) -2 = (-2)
```

```
6 - -7 =            7
6 + 7 = +         + 6
(Sign is positive.) 13 = (+13)
```

S1. -7 - 6 =

S2. -7 - -8 =

1. 3 - -12 =

2. 16 - 19 =

3. -14 - -22 =

4. -17 - 14 =

5. 50 - -16 =

6. 48 - 14 =

7. -8 - 12 =

8. -72 - -54 =

9. -39 - 54

10. -63 - -94 =

1.	
2.	
3.	
4.	
5.	
6.	
7.	
8.	
9.	
10.	

Score

Problem Solving

At night the temperature was 44 degrees. By morning it had dropped 52 degrees. What was the temperature in the morning?

Review Exercises

1. -7 + 6 + -8 =   2. 16 + -22 + -10 + 6 =   3. 32 + -46 + -16 =

4. -35 - 7 =   5. 16 - -4 =   6. 32 - 48 =

Helpful Hints

Use what you have learned to solve the following problems.
Remember, to subtract means to add its opposite.

Examples: -8 - 20 = 8 + -20   -8 - -20 = -8 + 20

S1. -12 - 16 =	S2. 50 - -62 =	1. -22 - -60 =	1.
			2.
			3.
2. 15 - 23 =	3. -24 - -36 =	4. -14 - 32 =	4.
			5.
			6.
5. 55 - -16 =	6. -39 - 40 =	7. -6 - 5 - 3 =	7.
			8.
			9.
8. -102 - -150 =	9. -220 - 214 =	10. -58 - -42 =	10.
			Score

Problem Solving

Together, Juan and James have earned 520 dollars.
If Juan has earned 325 dollars, how much has James earned?

Review Exercises

1. -7 - 9 =

2. -12 - -10 =

3. 18 - -22 =

4. 16 + -7 + 12 =

5. -8 + 6 + -7 + 5 =

6. -22 + -72 + -13 =

Helpful Hints	Use what you have learned to solve the problems on this page. **Examples:**

$-7 + 4 + -3 + 2 =$ 10
$-10 + 6 = -$ $- 6$
(Sign is negative.) $\overline{4} = \boxed{-4}$

$-7 - -6 =$ 7
$-7 + 6 = -$ $- 6$
(Sign is negative.) $\overline{1} = \boxed{-1}$

$15 - 36 =$ 36
$15 + -36 = -$ $- 15$
(Sign is negative.) $\overline{21} = \boxed{-21}$

S1. -94 + 48 =

S2. 15 - -7 =

1. -42 + -63 =

2. -95 + 110 =

3. -9 - 12 =

4. 40 - -20 =

5. -12 + 5 + -16 =

6. 20 + -7 + 4 + -8 =

7. 64 - 93 =

8. 7 - -12 =

9. -425 + 501 =

10. -723 - 201 =

1.

2.

3.

4.

5.

6.

7.

8.

9.

10.

Score

Problem Solving

If the temperature was -12° at midnight and by 6:00 A.M. it had dropped another 22°, what was the temperature at 6:00 A.M.?

Review Exercises

1. 336 + -521 =

2. -75 - 96 =

3. -402 + 763 =

4. 91 - -65 =

5. -256 + -758 =

6. -7 - 9 - 6 =

Helpful Hints

Use what you have learned to solve the following problems.
If you need help, refer to the examples on the previous page.

S1. -763 - 202 =

S2. 95 - -62 =

1. -29 - 36 =

2. -428 + 500 =

3. 50 - -21 =

4. 72 - 125 =

5. 65 + 12 + -52 + 16 =

6. -37 + -16 + -42 =

7. -95 - 24 =

8. -55 - -30 =

9. -33 - 35 =

10. -316 + -422 =

1.

2.

3.

4.

5.

6.

7.

8.

9.

10.

Score

Problem Solving

If the temperature was -12° at 6:00 A.M. and by noon
it had risen 33°, what was the temperature at noon?

Review Exercises

1. 337
    98
    324
    + 7

2. 1,712
    - 963

3. 304
    x 27

4. 63⟌1724

5. -27 - 37 =

6. 27 - -37 =

Helpful Hints	The product of two integers with different signs is negative. The product of two integers with the same sign is positive. (• means multiply).

Examples:

7 • -16 = -
(Sign is negative.)
$\begin{array}{r} 16 \\ \times\ 7 \\ \hline 112 \end{array}$ = (-112)

-8 • -7 = +
(Sign is positive.)
$\begin{array}{r} 8 \\ \times\ 7 \\ \hline 56 \end{array}$ = (+56)

S1. -4 x -12 =

S2. -17 • 8 =

1. -6 • -29 =

2. 16 • -5 =

3. -36 • -14 =

4. 27 • -22 =

5. -40 • 36 =

6. 9 x -19 =

7. -8 • -7 =

8. -24 • -16 =

9. 34 x -8 =

10. -17 • -38 =

1.

2.

3.

4.

5.

6.

7.

8.

9.

10.

Score

Problem Solving	Bill, Robert, and Olga together earned 520 dollars on Monday and 470 dollars on Tuesday. If they wanted to divide the money equally, how much would each person get?

Review Exercises

1. -6 • -7 =

2. 12 x -13 =

3. -15 • -20 =

4. 63 - 75 =

5. -66 + 95 =

6. -75 - -90 =

Helpful Hints	Use what you have learned to solve the following problems.
	Remember: The product of two integers with different signs is negative.
	The product of two integers with the same signs is positive.

S1. -15 • 12 =

S2. -16 • -20 =

1. -5 • -22 =

2. 5 x -22 =

3. -36 x -12 =

4. 42 x -10 =

5. -320 x 5 =

6. -75 • -5 =

7. 6 • -220 =

8. -18 • -40 =

9. 32 • -18 =

10. -160 • -15 =

1.

2.

3.

4.

5.

6.

7.

8.

9.

10.

Score

Problem Solving	A rope is 525 feet long. If it is cut into 5 pieces of equal length, how long will each piece be?

Review Exercises

1. -9 + 6 + -12 + 8 =

2. -7 - -9 =

3. -6 • -7 =

4. -6 x -42 =

5. -7 + -9 + -7 =

6. 16 - -20 =

Helpful Hints

When multiplying more than two integers, group them in pairs to simplify.

An integer next to parentheses means to multiply.

Examples:

2 • -3 (-6) =
(2 • -3) (-6) =
-6 (-6) = +
(Sign is positive.)

$$\begin{array}{r} 6 \\ \times\ 6 \\ \hline 36 \end{array} = \boxed{+36}$$

-2 • -3 • 4 • -2 =
(-2 • -3) • (4 • -2) =
6 • -8 = -
(Sign is negative.)

$$\begin{array}{r} 8 \\ \times\ 6 \\ \hline 48 \end{array} = \boxed{-48}$$

S1. -2 • 8 • -4 =

S2. (-2)(-5) • -3 =

1. 4 (-3) • 5 =

2. -3 • -6 (-7) =

3. 3 • -4 • -5 • 6 =

4. 5 (3) • -2 x (-6) =

5. 2 • -3 • -1 • -2 =

6. (-6) (-4) (-5) =

7. -3 (-4) • 1 (-4) =

8. 4 (-6) • 2 (-4) =

9. (-6) (-7) (2) (3) =

10. 11 (-10) (-4) =

1.	
2.	
3.	
4.	
5.	
6.	
7.	
8.	
9.	
10.	
Score	

Problem Solving

If a plane can travel 460 miles per hour, how far can it travel in 8 hours?

Review Exercises

1. -72 + 96 =

2. -53 - 52 =

3. 56 + -62 - -50 =

4. 7 · -12 =

5. -2 (-6) (-3) =

6. (-2) (-3) (-4) =

Helpful Hints	Use what you have learned to solve the following problems. Remember, when multiplying more than two integers, group them into pairs to simplify.

S1. -9 · 6 · -7 =	S2. -10 (-6) · (-3) =	1. 4 (-5) (-6) =	1.
			2.
			3.
2. -7 · -3 (-5) =	3. -2 · -3 · 4 · -6 =	4. 5 (6) · -5 · (-3) =	4.
			5.
5. (-5) (-8) (-5) =	6. (2) (-3) (4) (-5) =	7. (-7) (-4) · 2 (-6) =	6.
			7.
			8.
			9.
8. 5 (-6) · 4 (-3) =	9. (-6) (7) (-3) (2) =	10. -12 (3) (-6) =	10.
			Score

Problem Solving	A farmer owns 750 cows. If he decides to sell one-half of them, how many cows will he sell?

Review Exercises

1. -7 + 12 + -6 =

2. -7 - -9 =

3. 3 + -7 - -9 =

4. -63 - 72 - -60 =

5. 3 (-3) (-2) =

6. -5 (2) (2) (-3) =

Helpful Hints

The quotient of two integers with different signs is negative.

The quotient of two integers with the same sign is positive. (Hint: determine the sign, then divide.).

Examples:

$36 \div -4 = -$
(Sign is negative.)

$$4\overline{)36} \atop \underline{-36} \atop 0} \overset{9}{} = \boxed{-9}$$

$\dfrac{-123}{-2} = +$
(Sign is positive.)

$$3\overline{)123} \atop \underline{-12\downarrow} \atop 3} \overset{41}{} = \boxed{+4}$$

S1. 12 ÷ -4 =

S2. $\dfrac{-75}{-5}$

1. -72 ÷ 4 =

2. 336 ÷ -7 =

3. $\dfrac{-105}{-5}$

4. 204 ÷ -4 =

5. $\dfrac{-130}{5}$

6. 576 ÷ -12 =

7. 56 ÷ -7 =

8. 357 ÷ -21 =

9. $\dfrac{-65}{-5}$

10. -684 ÷ -36 =

1.
2.
3.
4.
5.
6.
7.
8.
9.
10.
Score

Problem Solving

Eva is inviting 70 people to a party. She plans to provide one soft drink for each person invited. If soft drinks come in packs of six, how many packs must she buy?

Review Exercises

1. $7\overline{)23}$ 2. $7\overline{)1396}$ 3. $60\overline{)896}$

4. $60\overline{)394}$ 5. $32\overline{)186}$ 6. $32\overline{)489}$

| **Helpful Hints** | Use what you have learned to solve the following problems. The quotient of two integers with different signs is negative. The quotient of two integers with the same sing is positive. |

S1. $96 \div -3 =$ S2. $\dfrac{-90}{-15}$ 1. $-72 \div 4 =$

2. $324 \div -4 =$ 3. $\dfrac{-115}{5}$ 4. $\dfrac{-2121}{-3}$

5. $864 \div -12 =$ 6. $\dfrac{-110}{-5}$ 7. $-80 \div -5 =$

8. $\dfrac{65}{-13}$ 9. $-104 \div 4 =$ 10. $-2001 \div -3 =$

1.	
2.	
3.	
4.	
5.	
6.	
7.	
8.	
9.	
10.	
Score	

| **Problem Solving** | A car traveled 295 miles in 5 hours. What was the car's average speed? |

63

Review Exercises

1. 32 (-3) =

2. (-2) (-6) =

3. (-3) 2 (-4) =

4. 200 ÷ -4 =

5. $\dfrac{60}{-15}$

6. -90 ÷ -15 =

Helpful Hints	Use what you have learned to solve problems like these.	**Examples:** $\dfrac{-36 \div -9}{4 \div -2} = \dfrac{4}{-2} = \boxed{-2}$ (Sign is negative.) $\dfrac{4 \times -8}{-8 \div 2} = \dfrac{-32}{-4} = \boxed{+8}$ (Sign is positive.)

S1. $\dfrac{-15 \div -3}{5 \div -1} =$

S2. $\dfrac{3 \cdot (-8)}{-8 \div -4} =$

1. $\dfrac{-32 \div 4}{2 \cdot -2} =$

2. $\dfrac{-6 \times -5}{-30 \div -3} =$

3. $\dfrac{4 \cdot (-6)}{(-2)(-3)} =$

4. $\dfrac{-4 \cdot -9}{8 \div -4} =$

5. $\dfrac{-36 \div -6}{-10 \div 5} =$

6. $\dfrac{-24 \div -3}{-2 \cdot -2} =$

7. $\dfrac{75 \div -25}{3 \div -1} =$

8. $\dfrac{-42 \div -2}{14 \div -2} =$

9. $\dfrac{45 \div 5}{-9 \div 3} =$

10. $\dfrac{-56 \div -7}{-36 \div -9} =$

1.

2.

3.

4.

5.

6.

7.

8.

9.

10.

Score

Problem Solving	Kale has a library book that is seven days overdue. For each day overdue he must pay 25 cents. How much must he pay the library for his overdue book?

Review Exercises

1. Find the difference between 709 and 688.

2. Find the product of 42 and 604.

3. Find the quotient of 612 and 3.

4. $-77 - -60 =$

5. $(-3)(-2)(2) =$

6. $\dfrac{36 \div -9}{-8 \div -4} =$

Helpful Hints	Use what you have learned to solve problems. If necessary, refer to the examples on the previous page.

S1. $\dfrac{32 \div -4}{-8 \div 2} =$

S2. $\dfrac{5 \cdot -6}{-12 \div -4} =$

1. $\dfrac{2 \cdot -4}{-12 \div -6} =$

2. $\dfrac{-8 \times -5}{15 \div -3} =$

3. $\dfrac{4 \cdot -9}{-18 \div -3} =$

4. $\dfrac{-4 \cdot -20}{-32 \div 4} =$

5. $\dfrac{-54 \div -9}{-2 \cdot -3} =$

6. $\dfrac{40 \div -5}{-20 \div 5} =$

7. $\dfrac{10 \cdot -3}{-3 \cdot -5} =$

8. $\dfrac{32 \div -2}{-2 \cdot -2} =$

9. $\dfrac{-55 \div 11}{-25 \div -5} =$

10. $\dfrac{-42 \div 7}{-12 \div -4} =$

1.

2.

3.

4.

5.

6.

7.

8.

9.

10.

Score

Problem Solving	Elena scored a total of 475 points on five tests. What was her average score?

Review Exercises

1. $77 + 99 + 73 + 62 =$

2. $\begin{array}{r} 7,165 \\ 769 \\ + \;\; 963 \\ \hline \end{array}$

3. $\begin{array}{r} 308 \\ \times \;\; 206 \\ \hline \end{array}$

4. $\begin{array}{r} 72,172 \\ - \;\; 15,564 \\ \hline \end{array}$

5. $6 \times 1,096 =$

6. $6,102 - 1,765 =$

Helpful Hints	Use what you have learned to solve the following problems. Remember to be careful with positive and negative signs.

S1. $(-5) \cdot 3 \,(-4) \times 6 =$ S2. $\dfrac{20 \div -2}{-25 \div 5} =$ 1. $15 \cdot -6 =$

2. $6 \times -5 \cdot 4 =$ 3. $(-3)\,(2)\,(-4) =$ 4. $\dfrac{-75}{-5}$

5. $32 \div -4 =$ 6. $-224 \div -4 =$ 7. $(-3)\,(-2)\,(6) =$

8. $\dfrac{-30 \div -5}{2 \cdot -3} =$ 9. $\dfrac{20 \times -3}{-50 \div -10} =$ 10. $\dfrac{36 \div -9}{-16 \div 4} =$

1.
2.
3.
4.
5.
6.
7.
8.
9.
10.
Score

Problem Solving	Fariba had test scores of 80, 84, 96, and 100. What was her average score?

Review Exercises

1. $3\overline{)76}$

2. $\dfrac{1250}{5}$

3. $41\overline{)361}$

4. $41\overline{)765}$

5. $28\overline{)349}$

6. $28\overline{)2106}$

Helpful Hints	Use what you have learned to solve the following problems. Sometimes it is helpful to review the examples from previous pages if you need help.

S1. $-3 \cdot 4 \cdot -4 \cdot 3 =$	S2. $\dfrac{15 \cdot -2}{-2 \cdot -3}$	1. $25 \cdot (-6) =$	1. _____
			2. _____
			3. _____
2. $5 \cdot (-6)(-3) =$	3. $(5)(-4)(2)(-2) =$	4. $\dfrac{-125}{-5}$	4. _____
			5. _____
			6. _____
5. $54 \div -6 =$	6. $-111 \div -3 =$	7. $\dfrac{(-3)(-6)}{-9}$	7. _____
			8. _____
			9. _____
8. $\dfrac{60 \div -6}{-2 \cdot -5} =$	9. $\dfrac{-72 \div 9}{-20 \div -5} =$	10. $\dfrac{20 \text{ x } -3}{-4 \text{ x } -5} =$	10. _____
			Score

Problem Solving	The attendance at the Eagle's game last year was 42,728. This year 53,962 attended. What was the increase in attendance this year?

Review of All Integer Operations

1. -8 + 5 =

2. 8 + -5 =

3. -8 + -5 =

4. -6 + -8 + 17 =

5. -35 + 19 + 23 + -33 =

6. 6 - 8 =

7. 4 - -8 =

8. -3 - 7 =

9. -12 - 16 =

10. 18 - 19 =

11. 4 • -15 =

12. -3 • -27 =

13. 3 (-6) (-4) =

14. (-2) • 5 (-6) • 2 =

15. -27 ÷ 9 =

16. -234 ÷ -3 =

17. $\dfrac{-256}{-8}$

18. $\dfrac{-24 \div 2}{18 \div 3}$

19. $\dfrac{8 \cdot (-4)}{-20 \div -5}$

20. $\dfrac{-30 \cdot -2}{-60 \div -10}$

1.	
2.	
3.	
4.	
5.	
6.	
7.	
8.	
9.	
10.	
11.	
12.	
13.	
14.	
15.	
16.	
17.	
18.	
19.	
20.	

Final Review of All Integer Operations

1. $12 + {}^-6 =$

2. $-12 + 6 =$

3. $-12 + {}^-6 =$

4. $-6 + {}^-9 + 21 =$

5. $65 + {}^-32 + 15 + {}^-50 =$

6. $9 - 12 =$

7. $15 - {}^-16 =$

8. $-20 - 31 =$

9. $-23 - 26 =$

10. $16 - 80 =$

11. $-16 \cdot 7 =$

12. $-15 \cdot {}^-10 =$

13. $3\,({}^-12)\,({}^-4) =$

14. $({}^-2)\,3\,({}^-5)\,3 =$

15. $\dfrac{{}^-72}{{}^-8}$

16. $-414 \div {}^-3 =$

17. $\dfrac{884}{{}^-4}$

18. $\dfrac{80 \div {}^-8}{{}^-25 \div 5}$

19. $\dfrac{{}^-12 \cdot {}^-3}{{}^-2 \cdot 2}$

20. $\dfrac{75 \div {}^-25}{{}^-3 \div {}^-1}$

| 1. |
| 2. |
| 3. |
| 4. |
| 5. |
| 6. |
| 7. |
| 8. |
| 9. |
| 10. |
| 11. |
| 12. |
| 13. |
| 14. |
| 15. |
| 16. |
| 17. |
| 18. |
| 19. |
| 20. |

Whole Numbers - Final Test

1.
```
   356
 + 397
```

2.
```
   623
   462
    17
 + 816
```

3. $6,502 + 916 + 15,989 =$

4. $6,667 + 3,444 + 755 + 16 =$

5. $7,093 + 675 + 97 + 768 =$

6.
```
   916
 - 428
```

7.
```
  5,392
 - 1,768
```

8. $7,053 - 4,289 =$

9. $5,000 - 3,296 =$

10. $7,008 - 999 =$

11.
```
    87
 x   4
```

12.
```
  7,132
 x    4
```

13.
```
    45
 x  36
```

14.
```
   392
 x  47
```

15.
```
   743
 x 247
```

16. $4\overline{)626}$

17. $4\overline{)1428}$

18. $40\overline{)568}$

19. $30\overline{)8672}$

20. $18\overline{)1343}$

1.
2.
3.
4.
5.
6.
7.
8.
9.
10.
11.
12.
13.
14.
15.
16.
17.
18.
19.
20.

Integers - Final Test

1. 7 + -5 =

2. -7 + 5 =

3. -7 + -5 =

4. -7 + 4 + -3 + 6 =

5. -52 + 17 + 22 + -21 =

6. 7 - 9 =

7. 4 - -7 =

8. -7 - 12 =

9. -12 - 16 =

10. 14 - 19 =

11. 4 • (-8) =

12. -12 • -19 =

13. 4 (-5) (-6) =

14. -2 • 3 • (-7) =

15. -63 ÷ 9 =

16. -560 ÷ -5 =

17. $\dfrac{-136}{-8}$

18. $\dfrac{32 \div -2}{-16 \div -4}$

19. $\dfrac{-8 \cdot -5}{-15 \div 3}$

20. $\dfrac{-40 \cdot -2}{-20 \div 2}$

1.
2.
3.
4.
5.
6.
7.
8.
9.
10.
11.
12.
13.
14.
15.
16.
17.
18.
19.
20.

Section 2

Fractions

Review Exercises

Note to the students and teachers: This section will include problems from all topics covered in this book.
Here are some simple problems to get started.

1. $35 + 21 + 16 =$ 2. $317 + 23 + 209 =$ 3.   63
                                                                729
                                                              + 472

4. $715 - 387 =$ 5.   621   6.   400
                              - 173   - 76

Helpful Hints	A fraction is a number that names a part of a whole or group.	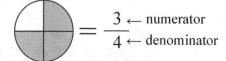 $= \dfrac{3}{4} \begin{matrix}\leftarrow \text{numerator} \\ \leftarrow \text{denominator}\end{matrix}$ Think of $\dfrac{3}{4}$ as $\dfrac{3 \text{ of}}{4 \text{ equal parts}}$

Write a fraction for each shaded part.
Then write a fraction for each unshaded (white) part.

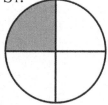

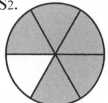

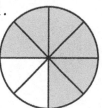

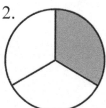

1.

2.

3.

4.

5.

6.

7.

8.

9.

10.

Score

Problem Solving	Crayons come in boxes of 24. How many crayons are there in fifteen boxes?

Review Exercises

1. 24 + 72 + 31 =

2. 705 - 76 =

3. 712
    - 176

4. 55
    666
    + 777

5. Find the difference
    of 752 and 317.

6. Find the sum of
    27, 29, 53, and 64.

| **Helpful Hints** | Use what you have learned to solve the following problems. * Some fractions may have more than one name. | Example: This shaded part can be written as $\frac{1}{2}$ and $\frac{2}{4}$. |

Write the fraction for each shaded part.
Then write a fraction for each unshaded (white) part.

S1.

S2.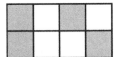

1.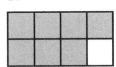

2.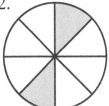

3.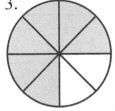

4.

5.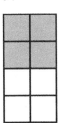

6.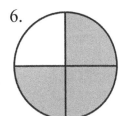

7.

8.

9.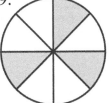

10.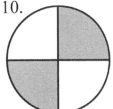

1.

2.

3.

4.

5.

6.

7.

8.

9.

10.

Score

| **Problem Solving** | Roger earned $850. If he spent $79 for groceries, how much of his earnings were left? |

Review Exercises

Write a fraction for each shaded part. Some may have more than one name.

1.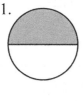

2.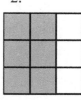

3.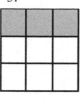

4.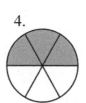

5.

6.

Helpful Hints	There are many ways to sketch a given fraction.	**Examples:** $\frac{1}{3} =$ 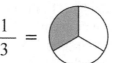 or or

Make a sketch for each fraction. Write the fraction in words in the answer column.

Example: $\frac{2}{5} =$ two fifths

S1. $\frac{1}{5}$ S2. $\frac{3}{8}$ 1. $\frac{3}{4}$ 2. $\frac{2}{9}$

3. $\frac{4}{5}$ 4. $\frac{4}{7}$ 5. $\frac{7}{12}$ 6. $\frac{7}{8}$

7. $\frac{4}{8}$ 8. $\frac{2}{3}$ 9. $\frac{5}{6}$ 10. $\frac{4}{6}$

1.

2.

3.

4.

5.

6.

7.

8.

9.

10.

Score

Problem Solving	A car traveled 55 miles per hour for eight hours. How far did the car travel?

Review Exercises

1.   127
      14
   + 316

2.   Sketch a figure for $\frac{5}{6}$

3.   Write a fraction for the shaded part.

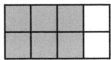

4.   501 - 36 =

5.   953
    - 271

6.   37 + 128 + 27 =

| **Helpful Hints** | Use what you have learned to solve the following problems. * Some fractions may have more than one name. | **Examples:** $\frac{3}{4}$ =  or or |

Make a sketch for each fraction. Then write the fraction in words in the answer column.

Example: $\frac{3}{8}$ = three eights

S1. $\frac{1}{3}$	S2. $\frac{3}{10}$
1. $\frac{5}{8}$	2. $\frac{1}{6}$

Answer column
1.
2.
3.
4.
5.
6.
7.
8.
9.
10.
Score

3. $\frac{1}{8}$ 4. $\frac{3}{12}$ 5. $\frac{2}{3}$ 6. $\frac{7}{9}$

7. $\frac{2}{7}$ 8. $\frac{6}{9}$ 9. $\frac{9}{10}$ 10. $\frac{2}{4}$

| **Problem Solving** | Maya is 16 years older than her sister. If Maya's sister is 11, how old is Maya? |

Review Exercises

1. Write two fractions for the shaded part.

2. Write two fractions for the shaded part.

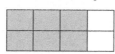

3. Sketch a figure for $\frac{2}{3}$.

4. $552 + 78 + 162 =$

5. $700 - 263 =$

6. Find the difference between 752 and 385.

Helpful Hints

$\frac{2}{4}$ has been reduced to its simplest form which is $\frac{1}{2}$. To do so, divide the numerator and denominator by the largest possible number.

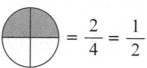 $= \frac{2}{4} = \frac{1}{2}$

Examples:

$2\overline{)\frac{6}{8}} = \frac{3}{4}$

Sometimes more than one step can be used:

$2\overline{)\frac{24}{28}} = 2\overline{)\frac{12}{14}} = \frac{6}{7}$

Reduce each fraction to its lowest terms.

S1. $\frac{5}{10}$

S2. $\frac{12}{16}$

1. $\frac{12}{15}$

2. $\frac{15}{20}$

3. $\frac{10}{20}$

4. $\frac{20}{25}$

5. $\frac{12}{18}$

6. $\frac{16}{24}$

7. $\frac{24}{40}$

8. $\frac{20}{32}$

9. $\frac{15}{18}$

10. $\frac{18}{24}$

1.	
2.	
3.	
4.	
5.	
6.	
7.	
8.	
9.	
10.	
Score	

Problem Solving

If a car can go 32 miles per gallon of gas, how many gallons will the car use traveling 384 miles?

Review Exercises

1. Reduce $\dfrac{6}{8}$ to its simplest form.

2. Reduce $\dfrac{20}{25}$ to its simplest form.

3. $755 - 666 =$

4. Sketch a figure for $\dfrac{7}{8}$.

5.
$$\begin{array}{r} 764 \\ 44 \\ + \ 555 \\ \hline \end{array}$$

6. Find the sum of 6,123 and 7,697.

Helpful Hints	Sometimes fractions can be reduced to simplest form in one step. However, it is okay to use multiple steps.	**Examples:** $12\overline{\smash{\big)}\dfrac{24}{36}} = \dfrac{2}{3}$ $2\overline{\smash{\big)}\dfrac{24}{36}} = 2\overline{\smash{\big)}\dfrac{12}{18}} = 2\overline{\smash{\big)}\dfrac{6}{9}} = \dfrac{2}{3}$

Reduce each fraction to its lowest terms.

S1. $\dfrac{15}{20}$ S2. $\dfrac{20}{24}$ 1. $\dfrac{12}{20}$ 2. $\dfrac{30}{40}$

3. $\dfrac{16}{30}$ 4. $\dfrac{14}{16}$ 5. $\dfrac{75}{100}$ 6. $\dfrac{30}{48}$

7. $\dfrac{30}{32}$ 8. $\dfrac{25}{30}$ 9. $\dfrac{80}{90}$ 10. $\dfrac{40}{48}$

1.	
2.	
3.	
4.	
5.	
6.	
7.	
8.	
9.	
10.	
Score	

Problem Solving

Juan planned a 3-day, 50 mile hike. If he hikes 17 miles the first day and 19 the second day, how far must he hike on the third day?

Review Exercises

1.   715
    - 367

2. $335 + 36 + 418 =$

3. $7000 - 763 =$

4.   Reduce $\dfrac{12}{16}$ to its simplest form.

5.   127
     14
   + 316

6.   Reduce $\dfrac{20}{24}$ to its simplest form.

| **Helpful Hints** | Equivalent fractions have the same value. Sometimes it is necessary to write a fraction as an equivalent fraction. Also, any fraction with the same numerator and denominator equals 1. We can multiply any fraction by 1 and not change its value. | **Examples:** $\dfrac{2}{2}$ or $\dfrac{5}{5}$ or $\dfrac{10}{10}$ all equal 1.
 Examples: $\dfrac{2}{3} \times \dfrac{5}{5} = \dfrac{10}{15}$ So, $\dfrac{2}{3} = \dfrac{10}{15}$. |

It is easy to solve the problem below. To find the missing numerator, multiply $\dfrac{2}{6}$ by $\dfrac{3}{3}$.

$$\dfrac{2}{6} = \dfrac{}{18} \qquad \dfrac{2}{6} \times \dfrac{3}{3} = \dfrac{6}{18} \qquad \text{So, } \dfrac{2}{6} = \dfrac{6}{18}. \qquad \text{The missing numerator is 6.}$$

Find the missing numerators.

S1. $\dfrac{3}{4} = \dfrac{}{20}$

S2. $\dfrac{3}{4} = \dfrac{}{12}$

1. $\dfrac{2}{7} = \dfrac{}{35}$

2. $\dfrac{3}{4} = \dfrac{}{16}$

3. $\dfrac{2}{3} = \dfrac{}{27}$

4. $\dfrac{11}{12} = \dfrac{}{24}$

5. $\dfrac{7}{10} = \dfrac{}{50}$

6. $\dfrac{9}{14} = \dfrac{}{28}$

7. $\dfrac{3}{13} = \dfrac{}{26}$

8. $\dfrac{2}{3} = \dfrac{}{60}$

9. $\dfrac{4}{9} = \dfrac{}{36}$

10. $\dfrac{11}{15} = \dfrac{}{45}$

1.
2.
3.
4.
5.
6.
7.
8.
9.
10.

Score

Problem Solving

A school spent $2,550.00 on 85 math books. How much did each book cost?

Review Exercises

1. Write a fraction for the shaded part.

2. $\begin{array}{r} 629 \\ -\ 348 \\ \hline \end{array}$

3. Write two fractions for the shaded portions.

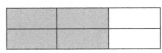

5. Find the missing numerator.

$$\frac{7}{8} = \frac{}{24}$$

4. $36 + 93 + 42 =$

6. $8,000 - 715 =$

Helpful Hints

Use what you have learned to solve the following problems.

Find the missing numerators.

S1. $\dfrac{9}{10} = \dfrac{}{20}$ S2. $\dfrac{15}{18} = \dfrac{}{36}$ 1. $\dfrac{4}{15} = \dfrac{}{45}$ 2. $\dfrac{11}{20} = \dfrac{}{80}$

3. $\dfrac{7}{8} = \dfrac{}{40}$ 4. $\dfrac{15}{16} = \dfrac{}{32}$ 5. $\dfrac{3}{7} = \dfrac{}{42}$ 6. $\dfrac{3}{13} = \dfrac{}{39}$

7. $\dfrac{2}{3} = \dfrac{}{9}$ 8. $\dfrac{11}{12} = \dfrac{}{36}$ 9. $\dfrac{4}{7} = \dfrac{}{56}$ 10. $\dfrac{7}{9} = \dfrac{}{63}$

1.
2.
3.
4.
5.
6.
7.
8.
9.
10.
Score

Problem Solving

A car traveled 78 miles on 2 gallons of gas.
How many miles per gallon did the car average?

Review Exercises

1. Write a fraction for the shaded part.

2. Reduce $\frac{6}{9}$ to its lowest terms.

3. Reduce $\frac{28}{50}$ to its lowest terms.

4.
```
  610
- 375
```

5.
```
  352
  463
  721
+ 445
```

6. Write three fractions for the shaded part.

Helpful Hints

An improper fraction has a numerator that is greater than or equal to its denominator. An improper fraction can be written either as a whole number or as a mixed numeral (a whole number and a fraction).

* Divide the numerator by the denominator.

$$2\overline{\smash)7} \quad \frac{3}{} = 3\frac{1}{2}$$
$$\frac{6}{1}$$

Example: $\bigcirc\bigcirc\bigcirc\ominus = \frac{7}{2} = 3\frac{1}{2}$

Change each improper fraction to a mixed numeral or whole number. Reduce answers to the lowest terms.

S1. $\frac{5}{4} =$

S2. $\frac{9}{6} =$

1. $\frac{11}{4} =$

2. $\frac{11}{7} =$

3. $\frac{45}{15} =$

4. $\frac{22}{4} =$

5. $\frac{37}{12} =$

6. $\frac{24}{6} =$

7. $\frac{55}{10} =$

8. $\frac{40}{6} =$

9. $\frac{18}{8} =$

10. $\frac{27}{5} =$

1.	
2.	
3.	
4.	
5.	
6.	
7.	
8.	
9.	
10.	

Problem Solving

On Saturday 12,034 people visited the museum. On Sunday there were 15,768 visitors. How many more people visited the museum on Sunday than on Saturday?

Score

Review Exercises

1. Change $\dfrac{11}{7}$ to a mixed numeral.

2. Change $\dfrac{27}{3}$ to a whole number.

3. Reduce $\dfrac{25}{30}$ to its lowest terms.

4. Give two names for the shaded part.

5. Find the sum of 2,563 and 3,699.

6. Find the difference between 9,112 and 7,132.

Helpful Hints	Use what you have learned to solve the following problems. For each mixed number, be sure that the fraction is reduced to its lowest terms.

Change each improper fraction to a mixed numeral or whole number. Reduce answers to the lowest terms.

S1. $\dfrac{19}{15} =$ S2. $\dfrac{65}{20} =$ 1. $\dfrac{20}{8} =$ 2. $\dfrac{9}{8} =$

3. $\dfrac{27}{5} =$ 4. $\dfrac{32}{6} =$ 5. $\dfrac{25}{4} =$ 6. $\dfrac{30}{12} =$

7. $\dfrac{85}{10} =$ 8. $\dfrac{16}{9} =$ 9. $\dfrac{96}{15} =$ 10. $\dfrac{40}{24} =$

1.

2.

3.

4.

5.

6.

7.

8.

9.

10.

Score

Problem Solving	A family of 2 adults and 4 children went to the movies. If adult's tickets are $9 each and children's tickets are $6 each, how much was the total cost of the tickets?

Review Exercises

1. Change $\frac{9}{5}$ to a mixed numeral.

2. Reduce $\frac{16}{20}$ to its lowest terms.

3. Change $\frac{40}{18}$ to a mixed numeral.

4. Find the missing numerator.

$$\frac{7}{8} = \frac{}{24}$$

5. Write two fractions for the shaded part.

6. Find the missing numerator.

$$\frac{}{12} = \frac{2}{3}$$

Helpful Hints

To change a mixed number to an improper fraction, do the following:
1. Multiply the denominator and the whole number.
2. Add the answer to the numerator.

Example:

$$3\frac{2}{5} = \frac{(15+2)}{5} = \frac{11}{7}$$

1. Multiply 2. Add

Change each mixed numeral to an improper fraction.

S1. $2\frac{1}{2} =$ S2. $6\frac{1}{3} =$ 1. $7\frac{1}{5} =$ 2. $3\frac{2}{3} =$

3. $5\frac{1}{4} =$ 4. $4\frac{3}{4} =$ 5. $2\frac{5}{6} =$ 6. $7\frac{1}{8} =$

7. $7\frac{2}{3} =$ 8. $8\frac{3}{4} =$ 9. $9\frac{1}{4} =$ 10. $1\frac{7}{8} =$

1.
2.
3.
4.
5.
6.
7.
8.
9.
10.
Score

Problem Solving

Lucy plans to drive 360 miles. If her car can travel 20 miles per gallon of gas, how many gallons of gas will she need to make the trip?

Review Exercises

1. Find the missing numerator.

$$\frac{11}{20} = \frac{}{60}$$

2. Find the missing numerator.

$$\frac{}{12} = \frac{3}{4}$$

3. Change $\frac{15}{7}$ to a mixed numeral.

4. Reduce $\frac{20}{25}$ to its lowest terms.

5. Draw a sketch for $\frac{5}{6}$.

6. Reduce $\frac{20}{30}$ to its lowest terms.

Helpful Hints	Use what you have learned to solve the following problems.

Change each mixed numeral to an improper fraction.

S1. $8\frac{1}{2} =$ S2. $4\frac{3}{4} =$ 1. $7\frac{1}{2} =$ 2. $5\frac{2}{3} =$

3. $5\frac{3}{8} =$ 4. $2\frac{2}{3} =$ 5. $9\frac{1}{3} =$ 6. $7\frac{3}{5} =$

7. $7\frac{2}{3} =$ 8. $16\frac{1}{2} =$ 9. $4\frac{3}{5} =$ 10. $5\frac{3}{4} =$

1.

2.

3.

4.

5.

6.

7.

8.

9.

10.

Score

Problem Solving	Maria needs 32 cupcakes for a party. If cupcakes come in packages of 6, how many packages must Maria buy?

Review Exercises

1. Change $\dfrac{10}{7}$ to a mixed numeral.

2. Change $4\dfrac{3}{5}$ to an improper fraction.

3. Reduce $\dfrac{25}{35}$ to its lowest terms.

4. Change $5\dfrac{1}{4}$ to an improper fraction.

5. Change $\dfrac{20}{15}$ to a mixed numeral.

6. Draw a sketch for $\dfrac{2}{3}$.

Helpful Hints	To add fractions with like denominators, add the numerators and then ask the following questions: 1. Is the fraction improper? If it is, make it a mixed numeral or whole number. 2. Can the fraction be reduced? If it can, reduce it to its simplest form.	**Example:** $\begin{aligned}&\dfrac{7}{10}\\+&\dfrac{5}{10}\\\hline&\dfrac{12}{10}=1\dfrac{2}{10}=1\dfrac{1}{5}\end{aligned}$

S1. $\dfrac{1}{10}$ $+\dfrac{7}{10}$

S2. $\dfrac{7}{8}$ $+\dfrac{3}{8}$

1. $\dfrac{5}{9}$ $+\dfrac{2}{9}$

2. $\dfrac{1}{8}$ $+\dfrac{5}{8}$

3. $\dfrac{7}{8}$ $+\dfrac{5}{8}$

4. $\dfrac{7}{10}$ $+\dfrac{2}{10}$

5. $\dfrac{5}{8}$ $+\dfrac{5}{8}$

6. $\dfrac{9}{16}$ $+\dfrac{11}{16}$

7. $\dfrac{2}{5}$ $\dfrac{4}{5}$ $+\dfrac{3}{5}$

8. $\dfrac{3}{8}$ $\dfrac{5}{8}$ $+\dfrac{2}{8}$

9. $\dfrac{9}{10}$ $+\dfrac{7}{10}$

10. $\dfrac{5}{6}$ $+\dfrac{5}{6}$

1. _____
2. _____
3. _____
4. _____
5. _____
6. _____
7. _____
8. _____
9. _____
10. _____

Problem Solving	A recipe calls for $\dfrac{3}{8}$ cup of flour and $\dfrac{7}{8}$ cup of sugar. How much flour and sugar is needed altogether?	Score

Review Exercises

1. Change $5\frac{1}{2}$ to an improper fraction.

2. Change $\frac{16}{5}$ to a mixed numeral.

3.
$$\frac{3}{5}$$
$$+\frac{1}{5}$$

4.
$$\frac{3}{4}$$
$$+\frac{3}{4}$$

5. Make a sketch for $\frac{7}{8}$.

6. Reduce $\frac{18}{20}$ to its lowest terms.

Helpful Hints

Use what you have learned to solve the following problems.
* Remember:
1. Is the fraction improper? If it is, make it a mixed numeral or whole number.
2. Can the fraction be reduced? If it can, reduce it to its simplest form.

Change each mixed number to an improper fraction.

1.	
2.	
3.	
4.	
5.	
6.	
7.	
8.	
9.	
10.	
Score	

S1.
$$\frac{7}{10}$$
$$+\frac{1}{10}$$

S2.
$$\frac{11}{12}$$
$$+\frac{7}{12}$$

1.
$$\frac{1}{8}$$
$$+\frac{5}{8}$$

2.
$$\frac{3}{10}$$
$$+\frac{4}{10}$$

3.
$$\frac{9}{10}$$
$$+\frac{5}{10}$$

4.
$$\frac{11}{16}$$
$$+\frac{7}{16}$$

5.
$$\frac{7}{12}$$
$$+\frac{6}{12}$$

6.
$$\frac{3}{4}$$
$$\frac{1}{4}$$
$$+\frac{3}{4}$$

7.
$$\frac{5}{6}$$
$$+\frac{5}{6}$$

8.
$$\frac{9}{13}$$
$$+\frac{7}{13}$$

9.
$$\frac{2}{3}$$
$$\frac{2}{3}$$
$$+\frac{2}{3}$$

10.
$$\frac{11}{15}$$
$$+\frac{7}{15}$$

Problem Solving

Saturday it rained $\frac{7}{10}$ inches and Sunday it rained $\frac{9}{10}$ inches. What is the total amount of rain that fell during the two days?

Review Exercises

1. $\dfrac{5}{8}$
   $+\dfrac{4}{8}$

2. $\dfrac{3}{10}$
   $+\dfrac{3}{10}$

3. Reduce $\dfrac{16}{18}$ to its lowest terms.

6. $\dfrac{3}{4}$
   $+\dfrac{1}{4}$

4. Change $\dfrac{21}{4}$ to a mixed numeral.

5. Change $3\dfrac{3}{4}$ to an improper fraction.

Helpful Hints

1. Add the fractions first.
2. Add the whole numbers next.
3. If there is an improper fraction, change it to a mixed numeral.
4. Add the mixed numeral to the whole number.

Example:

$6\dfrac{7}{8}$
$+2\dfrac{5}{8}$

* Reduce the answer to its lowest terms.

$8\dfrac{12}{8} = 8 + 1\dfrac{4}{8} = 9\dfrac{4}{8} = 9\dfrac{1}{2}$

S1. $4\dfrac{1}{10}$
    $+2\dfrac{7}{10}$

S2. $3\dfrac{7}{8}$
    $+4\dfrac{3}{8}$

1. $5\dfrac{4}{7}$
   $+2\dfrac{2}{7}$

2. $3\dfrac{7}{10}$
   $+2\dfrac{5}{10}$

3. $5\dfrac{3}{4}$
   $+6\dfrac{1}{4}$

4. $3\dfrac{5}{6}$
   $+2\dfrac{3}{6}$

5. $4\dfrac{5}{7}$
   $+2\dfrac{4}{7}$

6. $2\dfrac{3}{4}$
   $+4\dfrac{3}{4}$

7. $5\dfrac{9}{10}$
   $+4\dfrac{3}{10}$

8. $2\dfrac{11}{15}$
   $+3\dfrac{6}{15}$

9. $6\dfrac{3}{10}$
   $+3\dfrac{7}{10}$

10. $9\dfrac{3}{5}$
    $+2\dfrac{3}{5}$

1.	
2.	
3.	
4.	
5.	
6.	
7.	
8.	
9.	
10.	
Score	

Problem Solving

A baker uses $\dfrac{3}{4}$ cups of flour for each pie he bakes. He uses $\dfrac{1}{4}$ cups of flour for each cake. How much flour is used to make two pies and one cake?

Review Exercises

1. $\dfrac{1}{3}$
$+ \dfrac{1}{3}$

2. $\dfrac{7}{10}$
$+ \dfrac{1}{10}$

3. $\dfrac{14}{15}$
$+ \dfrac{4}{15}$

4. Reduce $\dfrac{30}{32}$ to its lowest terms.

5. Change $3\dfrac{5}{6}$ to an improper fraction.

6. Change $\dfrac{17}{3}$ to a mixed numeral.

Helpful Hints	Use what you have learned to solve the following problems. * Remember: 1. If there is an improper fraction, change it to a mixed numeral. 2. Reduce each fraction to its lowest terms.

Change each mixed number to an improper fraction.

S1. $3\dfrac{7}{10}$
$+ 4\dfrac{1}{10}$

S2. $5\dfrac{7}{10}$
$+ 3\dfrac{9}{10}$

1. $3\dfrac{1}{5}$
$+ 2\dfrac{3}{5}$

2. $4\dfrac{4}{5}$
$+ 3\dfrac{3}{5}$

3. $9\dfrac{1}{6}$
$+ 5\dfrac{5}{6}$

4. $7\dfrac{7}{8}$
$+ 5\dfrac{5}{8}$

5. $13\dfrac{3}{5}$
$+ 16\dfrac{3}{5}$

6. $9\dfrac{2}{3}$
$+ 7\dfrac{2}{3}$

7. $3\dfrac{7}{15}$
$+ 4\dfrac{9}{15}$

8. $2\dfrac{3}{16}$
$+ 3\dfrac{5}{16}$

9. $7\dfrac{7}{10}$
$+ 3\dfrac{3}{10}$

10. $7\dfrac{7}{11}$
$+ 5\dfrac{9}{11}$

1.

2.

3.

4.

5.

6.

7.

8.

9.

10.

Score

Problem Solving	Anna rides her bike $\dfrac{7}{8}$ miles to school. If she has already ridden $\dfrac{3}{8}$ miles, how much farther must she ride before she gets to school?

Review Exercises

1. Change $\frac{18}{12}$ to a mixed numeral.

2. Reduce $\frac{20}{25}$ to its lowest terms.

3. $\begin{array}{r} \frac{3}{5} \\ + \frac{4}{5} \\ \hline \end{array}$

4. $\begin{array}{r} \frac{7}{8} \\ + \frac{3}{8} \\ \hline \end{array}$

5. $\begin{array}{r} \frac{3}{10} \\ + \frac{3}{10} \\ \hline \end{array}$

6. $\begin{array}{r} 5\frac{6}{7} \\ + 3\frac{5}{7} \\ \hline \end{array}$

Helpful Hints	Use what you have learned to solve the following problems. * Remember: 1. Change improper fractions to a mixed numerals. 2. Reduce each fraction to its lowest terms.

S1. $\begin{array}{r} \frac{7}{8} \\ + \frac{3}{8} \\ \hline \end{array}$

S2. $\begin{array}{r} 4\frac{4}{5} \\ + 3\frac{2}{5} \\ \hline \end{array}$

1. $\begin{array}{r} \frac{3}{6} \\ + \frac{1}{6} \\ \hline \end{array}$

2. $\begin{array}{r} 4\frac{5}{8} \\ + 3\frac{3}{8} \\ \hline \end{array}$

3. $\begin{array}{r} \frac{9}{10} \\ + \frac{7}{10} \\ \hline \end{array}$

4. $\begin{array}{r} 5\frac{2}{3} \\ + 3\frac{2}{3} \\ \hline \end{array}$

5. $\begin{array}{r} 7\frac{7}{8} \\ + 3\frac{5}{8} \\ \hline \end{array}$

6. $\begin{array}{r} \frac{2}{3} \\ \frac{1}{3} \\ + \frac{2}{3} \\ \hline \end{array}$

7. $\begin{array}{r} \frac{5}{16} \\ + \frac{5}{16} \\ \hline \end{array}$

8. $\begin{array}{r} 9\frac{7}{9} \\ + 3\frac{2}{9} \\ \hline \end{array}$

9. $\begin{array}{r} \frac{5}{8} \\ + \frac{5}{8} \\ \hline \end{array}$

10. $\begin{array}{r} 2\frac{11}{12} \\ + 3\frac{5}{12} \\ \hline \end{array}$

1.
2.
3.
4.
5.
6.
7.
8.
9.
10.

Score

Problem Solving	Pablo worked $7\frac{3}{4}$ hours on Monday and $6\frac{3}{4}$ hours on Tuesday. How many hours did he work altogether?

Review Exercises

1. Draw a sketch for $\dfrac{5}{6}$.

2. Change $\dfrac{7}{2}$ to a mixed numeral.

3. Reduce $\dfrac{14}{20}$ to its lowest terms.

4. $\begin{array}{r} \dfrac{5}{6} \\ + \dfrac{3}{6} \\ \hline \end{array}$

5. $\begin{array}{r} \dfrac{9}{10} \\ + \dfrac{1}{10} \\ \hline \end{array}$

6. $\begin{array}{r} 7\dfrac{3}{4} \\ + 8\dfrac{3}{4} \\ \hline \end{array}$

Helpful Hints	Use what you have learned to solve the following problems.

S1. $\begin{array}{r} 4\dfrac{5}{7} \\ + 3\dfrac{1}{7} \\ \hline \end{array}$

S2. $\begin{array}{r} \dfrac{11}{12} \\ + \dfrac{7}{12} \\ \hline \end{array}$

1. $\begin{array}{r} \dfrac{3}{8} \\ + \dfrac{1}{8} \\ \hline \end{array}$

2. $\begin{array}{r} 5\dfrac{5}{8} \\ + 3\dfrac{5}{8} \\ \hline \end{array}$

3. $\begin{array}{r} \dfrac{7}{8} \\ \dfrac{3}{8} \\ + \dfrac{5}{8} \\ \hline \end{array}$

4. $\begin{array}{r} 9\dfrac{2}{3} \\ + 3\dfrac{2}{3} \\ \hline \end{array}$

5. $\begin{array}{r} \dfrac{7}{10} \\ + \dfrac{1}{10} \\ \hline \end{array}$

6. $\begin{array}{r} 6\dfrac{7}{10} \\ + 3\dfrac{5}{10} \\ \hline \end{array}$

7. $\begin{array}{r} \dfrac{11}{12} \\ + \dfrac{3}{12} \\ \hline \end{array}$

8. $\begin{array}{r} 4\dfrac{2}{3} \\ 3\dfrac{1}{3} \\ + 5\dfrac{2}{3} \\ \hline \end{array}$

9. $\begin{array}{r} 3\dfrac{3}{8} \\ 4\dfrac{7}{8} \\ + 3\dfrac{5}{8} \\ \hline \end{array}$

10. $\begin{array}{r} \dfrac{7}{9} \\ + \dfrac{3}{9} \\ \hline \end{array}$

1.

2.

3.

4.

5.

6.

7.

8.

9.

10.

Score

Problem Solving	Pierre scores of 86, 88, and 96 on his tests. What was his average score?

Review Exercises

1. $\frac{3}{8}$
    $+\ \frac{1}{8}$

2. $\frac{5}{7}$
    $+\ \frac{4}{7}$

3. $\frac{9}{16}$
    $+\ \frac{11}{16}$

4. $3\frac{1}{4}$
    $+\ 4\frac{3}{4}$

5. $8\frac{3}{10}$
    $+\ 7\frac{5}{10}$

6. $4\frac{3}{4}$
    $+\ 5\frac{3}{4}$

Helpful Hints	To subtract fractions that have like denominators, subtract the numerators. Reduce the fractions to their lowest terms.	**Example:** $\begin{array}{r}\frac{9}{10}\\ -\ \frac{3}{10}\\ \hline \frac{6}{10}=\frac{3}{5}\end{array}$

S1. $\frac{3}{4}$
     $-\ \frac{1}{4}$

S2. $\frac{11}{16}$
     $-\ \frac{1}{16}$

1. $\frac{7}{10}$
    $-\ \frac{1}{10}$

2. $\frac{4}{5}$
    $-\ \frac{1}{5}$

3. $\frac{7}{8}$
    $-\ \frac{1}{8}$

4. $\frac{17}{20}$
    $-\ \frac{2}{20}$

5. $\frac{17}{18}$
    $-\ \frac{3}{18}$

6. $\frac{11}{12}$
    $-\ \frac{5}{12}$

7. $\frac{24}{25}$
    $-\ \frac{4}{25}$

8. $\frac{11}{35}$
    $-\ \frac{1}{35}$

9. $\frac{11}{12}-\frac{2}{12}=$

10. $\frac{3}{10}$
    $-\ \frac{1}{10}$

1.

2.

3.

4.

5.

6.

7.

8.

9.

10.

Score

Problem Solving	Mrs. Lopez had $\frac{7}{8}$ pounds of sugar. If she used $\frac{3}{8}$ pounds in baking a cake, how many pounds of sugar were left?

Review Exercises

1. Reduce $\dfrac{20}{25}$ to its lowest terms.

2. $3\dfrac{3}{5}$
   $+\ 5\dfrac{4}{5}$
   $\overline{\phantom{+\ 5\dfrac{4}{5}}}$

3. Change $7\dfrac{2}{3}$ to an improper fraction.

4. $\dfrac{9}{10}$
   $-\ \dfrac{1}{10}$
   $\overline{\phantom{-\ \dfrac{1}{10}}}$

5. $\dfrac{4}{5}+\dfrac{3}{5}=$

6. Change $\dfrac{19}{7}$ to a mixed numeral.

Helpful Hints	Use what you have learned to solve the following problems. Remember to reduce answers to lowest terms.

S1. $\dfrac{7}{12}$
   $-\ \dfrac{1}{12}$

S2. $\dfrac{29}{32}$
   $-\ \dfrac{1}{32}$

1. $\dfrac{15}{16}$
   $-\ \dfrac{3}{16}$

2. $\dfrac{14}{15}$
   $-\ \dfrac{2}{15}$

3. $\dfrac{7}{8}-\dfrac{5}{8}=$

4. $\dfrac{11}{24}$
   $-\ \dfrac{3}{24}$

5. $\dfrac{19}{20}$
   $-\ \dfrac{4}{20}$

6. $\dfrac{11}{18}$
   $-\ \dfrac{5}{18}$

7. $\dfrac{2}{3}$
   $-\ \dfrac{1}{3}$

8. $\dfrac{23}{35}$
   $-\ \dfrac{2}{35}$

9. $\dfrac{49}{50}$
   $-\ \dfrac{9}{50}$

10. $\dfrac{13}{48}$
   $-\ \dfrac{1}{48}$

1.

2.

3.

4.

5.

6.

7.

8.

9.

10.

Score

Problem Solving	Fredo worked $3\dfrac{1}{4}$ hours on Monday, $2\dfrac{3}{4}$ hours on Tuesday, and $4\dfrac{3}{4}$ hours on Wednesday. How many hours did he work altogether?

Review Exercises

1. $\dfrac{3}{4}$
$-\dfrac{1}{4}$

2. $2\dfrac{2}{3}$
$+7\dfrac{1}{3}$

3. $\dfrac{7}{12}$
$-\dfrac{4}{12}$

4. $4\dfrac{5}{8}$
$+3\dfrac{3}{8}$

5. $12\dfrac{2}{3}$
$+13\dfrac{2}{3}$

6. $\dfrac{15}{16}$
$+\dfrac{11}{16}$

Helpful Hints

To subtract mixed numerals with like denominators, subtract the fractions first, then the whole numbers. Reduce fractions to lowest terms.

If the fraction can't be subtracted, take one from the whole number, increase the fraction, then subtract.

Examples:

$7\dfrac{3}{4}$
$-\;2\dfrac{1}{4}$
$6\dfrac{2}{4}=6\dfrac{1}{2}$

$6\cancel{7}\dfrac{1}{4}+\dfrac{4}{4}=\dfrac{5}{4}$
$-\;2\dfrac{3}{4}$
$4\dfrac{2}{4}=4\dfrac{1}{2}$

S1. $4\dfrac{3}{4}$
$-1\dfrac{1}{4}$

S2. $5\dfrac{1}{5}$
$-2\dfrac{3}{5}$

1. $6\dfrac{3}{8}$
$-2\dfrac{1}{8}$

2. $7\dfrac{1}{4}$
$-2\dfrac{3}{4}$

3. $8\dfrac{3}{8}$
$-4\dfrac{7}{8}$

4. $7\dfrac{7}{8}$
$-2\dfrac{1}{8}$

5. $4\dfrac{3}{10}$
$-1\dfrac{7}{10}$

6. $5\dfrac{3}{20}$
$-2\dfrac{11}{20}$

7. $5\dfrac{3}{5}$
$-2\dfrac{3}{5}$

8. $12\dfrac{4}{9}$
$-10\dfrac{7}{9}$

9. $7\dfrac{1}{7}$
$-2\dfrac{3}{7}$

10. $8\dfrac{1}{15}$
$-2\dfrac{7}{15}$

1.
2.
3.
4.
5.
6.
7.
8.
9.
10.
Score

Problem Solving

Jose and Hector are taking a $20\dfrac{1}{4}$ mile bike ride. If they have ridden $8\dfrac{3}{4}$ miles, how much farther do they have to ride?

Review Exercises

1. Change $\frac{19}{15}$ to a mixed numeral.

2. $\begin{array}{r} \frac{15}{16} \\ - \frac{3}{16} \\ \hline \end{array}$

3. Reduce $\frac{30}{32}$ to its lowest terms.

4. $\begin{array}{r} \frac{3}{7} \\ + \frac{5}{7} \\ \hline \end{array}$

5. Write two fractions for the shaded part.

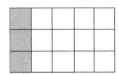

6. $\begin{array}{r} 4\frac{5}{8} \\ + 3\frac{5}{8} \\ \hline \end{array}$

Helpful Hints

Use what you have learned to solve the following problems.
* Remember:
1. If necessary take one whole from the whole number, increase the fraction, then subtract.
2. Reduce the answer to lowest terms.

	1.
	2.
	3.
	4.
	5.
	6.
	7.
	8.
	9.
	10.
	Score

S1. $\begin{array}{r} 5\frac{1}{4} \\ - 3\frac{3}{4} \\ \hline \end{array}$

S2. $\begin{array}{r} 9\frac{1}{8} \\ - 3\frac{5}{8} \\ \hline \end{array}$

1. $\begin{array}{r} 7\frac{2}{3} \\ - 1\frac{1}{3} \\ \hline \end{array}$

2. $\begin{array}{r} 8\frac{1}{3} \\ - 2\frac{2}{3} \\ \hline \end{array}$

3. $\begin{array}{r} 5\frac{7}{8} \\ - 1\frac{1}{8} \\ \hline \end{array}$

4. $\begin{array}{r} 6\frac{3}{8} \\ - 2\frac{7}{8} \\ \hline \end{array}$

5. $\begin{array}{r} 12\frac{7}{11} \\ - 9\frac{10}{11} \\ \hline \end{array}$

6. $\begin{array}{r} 7\frac{5}{12} \\ - 2\frac{7}{12} \\ \hline \end{array}$

7. $7\frac{3}{4} - 1\frac{1}{4} =$

8. $7\frac{1}{8} - 2\frac{3}{8} =$

9. $\begin{array}{r} 12\frac{1}{20} \\ - 7\frac{7}{20} \\ \hline \end{array}$

10. $\begin{array}{r} 7\frac{1}{9} \\ - 3\frac{7}{9} \\ \hline \end{array}$

Problem Solving

The Jones family traveled 80 miles in their car. If the car travels 20 miles per gallon of gas, how many gallons of gas did the car use? If gas is $3 per gallon, how much did the gas used for the trip cost?

Review Exercises

1. $\frac{7}{8}$
   $-\frac{1}{8}$

4. $\frac{5}{8}$
   $+\frac{7}{8}$

2. $3\frac{19}{20}$
   $-1\frac{4}{20}$

5. $3\frac{7}{10}$
   $+3\frac{1}{10}$

3. $7\frac{5}{8}$
   $-3\frac{7}{8}$

6. $5\frac{7}{15}$
   $+3\frac{10}{15}$

Helpful Hints

To subtract a fraction or mixed numeral from a whole number, take one from the whole number and make it a fraction, then subtract.

Examples:

$3\,\cancel{4} \rightarrow \frac{4}{4}$
$-2 \qquad \frac{1}{4}$
$\overline{\qquad 1\frac{3}{4}}$

$6\,\cancel{7} \rightarrow \frac{5}{5}$
$- \qquad \frac{3}{5}$
$\overline{\qquad 6\frac{2}{5}}$

S1. 7
    $-3\frac{3}{5}$

S2. 9
    $-\frac{5}{8}$

1. 12
   $-7\frac{1}{8}$

2. 6
   $-3\frac{3}{7}$

3. 12
   $-\frac{7}{15}$

4. 7
   $-2\frac{9}{10}$

5. 15
   $-2\frac{1}{8}$

6. 42
   $-29\frac{1}{7}$

7. 12
   $-\frac{9}{11}$

8. 13
   $-7\frac{1}{5}$

9. 53
   $-29\frac{3}{7}$

10. 16
    $-\frac{7}{15}$

1.
2.
3.
4.
5.
6.
7.
8.
9.
10.
Score

Problem Solving

A tailor had 15 yards of cloth. He used $5\frac{3}{4}$ yards to make a suit. How many yards of cloth were left?

Review Exercises

1. Make a sketch for $\frac{5}{6}$.

2.
$$7\frac{1}{3}$$
$$-\,2\frac{2}{3}$$

3. Change $\frac{23}{4}$ to a mixed numeral.

4.
$$\frac{7}{8}$$
$$+\;\frac{7}{8}$$

5.
$$\frac{9}{10}$$
$$-\;\frac{3}{10}$$

6. Change $3\frac{4}{5}$ to an improper fraction.

Helpful Hints

Use what you have learned to solve the following problems.

S1.
$$9$$
$$-\;\frac{7}{8}$$

S2.
$$12$$
$$-\,2\frac{3}{7}$$

1.
$$14$$
$$-\,9\frac{1}{4}$$

2.
$$61$$
$$-\,28\frac{3}{8}$$

3.
$$4$$
$$-\;\frac{15}{16}$$

4.
$$9$$
$$-\;\frac{15}{24}$$

5.
$$14$$
$$-\,3\frac{1}{24}$$

6.
$$22$$
$$-\,6\frac{9}{15}$$

7.
$$40$$
$$-\;\frac{13}{24}$$

8.
$$28$$
$$-\,19\frac{13}{24}$$

9.
$$15$$
$$-\,2\frac{19}{20}$$

10.
$$41$$
$$-\,16\frac{12}{15}$$

1.
2.
3.
4.
5.
6.
7.
8.
9.
10.
Score

Problem Solving

A car's gas tank contained 20 gallons of gas, On Monday $5\frac{1}{4}$ gallons were used, and on Tuesday $3\frac{1}{4}$ gallons were used. How many gallons of gas were left in he tank?

Review Exercises

1. $7\frac{1}{3}$

   $-2\frac{2}{3}$

2. $5\frac{7}{15}$

   $+3\frac{10}{15}$

3. Change $\frac{31}{25}$ to a mixed numeral.

4. Reduce $\frac{18}{30}$ to its lowest terms.

5. 7

   $-2\frac{3}{4}$

6. $\frac{11}{12} + \frac{3}{12} =$

Helpful Hints	Use what you have learned to solve the following problems. Regroup when necessary. Reduce all answers to their lowest terms. If problems are positioned horizontally, put them in columns before working.

1.
2.
3.
4.
5.
6.
7.
8.
9.
10.
Score

S1. $3\frac{9}{10}$

   $+2\frac{3}{10}$

S2. $6\frac{1}{5}$

   $-2\frac{3}{5}$

1. $\frac{6}{7}$

   $+\frac{3}{7}$

2. $\frac{14}{15}$

   $-\frac{4}{15}$

3. $7\frac{4}{9}$

   $+3\frac{7}{9}$

4. 9

   $-2\frac{1}{4}$

5. $7\frac{3}{8}$

   $+2\frac{3}{8}$

6. $9 - 2\frac{1}{2} =$

7. $4\frac{11}{12}$

   $+3\frac{5}{12}$

8. $7\frac{1}{3}$

   $-2\frac{2}{3}$

9. $12\frac{3}{10}$

   $-4\frac{7}{10}$

10. $\frac{9}{10}$

    $\frac{7}{10}$

    $+\frac{3}{10}$

Problem Solving	A plane traveled 2,700 miles in six hours. What was its average speed per hour?

Review Exercises

1. $5\frac{1}{4}$
   $-2\frac{3}{4}$

2. 7
   $-\frac{11}{15}$

3. $9\frac{4}{5}$
   $+7\frac{3}{5}$

4. Change $\frac{18}{12}$ to a mixed numeral.

5. Reduce $\frac{20}{32}$ to its lowest terms.

6. Change $5\frac{1}{4}$ to an improper fraction.

Helpful Hints	Use what you have learned to solve the following problems. Sometimes it is helpful to refer to the "Helpful Hints" section on previous pages.

S1. $7\frac{1}{8}$
    $-2\frac{3}{8}$

S2. $4\frac{5}{8}$
    $+3\frac{7}{8}$

1. $\frac{11}{12}$
   $-\frac{3}{12}$

2. $\frac{8}{9}$
   $+\frac{5}{9}$

3. $6\frac{3}{7}$
   $+4\frac{5}{7}$

4. 7
   $-\frac{12}{20}$

5. $23\frac{3}{8}$
   $+44\frac{7}{8}$

6. $7 - 3\frac{3}{4} =$

7. $9\frac{9}{10}$
   $+4\frac{3}{10}$

8. $6\frac{1}{12}$
   $-1\frac{7}{12}$

9. $3\frac{11}{15}$
   $+7\frac{4}{15}$

10. $\frac{7}{8}$
    $\frac{5}{8}$
    $+\frac{4}{8}$

1. _____
2. _____
3. _____
4. _____
5. _____
6. _____
7. _____
8. _____
9. _____
10. _____

Score _____

Problem Solving	There are 7 rows of desks in a classroom. There are 12 desks in each row. If 5 of the desks are taken, how many are empty?

Review Exercises

1. Find the sum of $\frac{7}{8}$ and $\frac{3}{8}$.

2. Find the difference between $3\frac{1}{4}$ and $1\frac{3}{4}$.

3. Reduce $\frac{42}{48}$ to its lowest terms.

4. Change $3\frac{2}{5}$ to an improper fraction.

5. $\frac{5}{6} + \frac{3}{6} + \frac{2}{6} =$

6. $7 - 2\frac{3}{4} =$

| **Helpful Hints** | To add or subtract fractions with unlike denominations you need to find the least common denominator (LCD). The LCD is the smallest number, other than zero, that each denominator will divide into evenly. | **Examples:** The least common denominator of: $\frac{1}{5}$ and $\frac{1}{10}$ is 10. $\frac{3}{8}$ and $\frac{1}{6}$ is 24. |

Find the least common denominator of each of the following.

S1. $\frac{3}{4}$ and $\frac{1}{5}$

S2. $\frac{5}{6}$ and $\frac{3}{8}$

1. $\frac{5}{6}$ and $\frac{1}{9}$

2. $\frac{3}{4}$ and $\frac{1}{12}$

3. $\frac{11}{21}$ and $\frac{1}{7}$

4. $\frac{1}{6}$, $\frac{1}{4}$ and $\frac{3}{8}$

5. $\frac{1}{9}$, $\frac{5}{6}$ and $\frac{7}{12}$

6. $\frac{3}{8}$ and $\frac{1}{7}$

7. $\frac{7}{12}$ and $\frac{5}{48}$

8. $\frac{1}{15}$, $\frac{9}{20}$ and $\frac{1}{6}$

9. $\frac{11}{12}$, $\frac{3}{8}$ and $\frac{11}{48}$

10. $\frac{3}{8}$, $\frac{3}{16}$ and $\frac{1}{12}$

1.

2.

3.

4.

5.

6.

7.

8.

9.

10.

Score

| **Problem Solving** | There are 468 students in a school. If they are divided in 18 equally sized classes, how many are in each class? |

Review Exercises

1. $\dfrac{4}{5}$
$+\dfrac{2}{5}$

2. $9\dfrac{5}{8}$
$+6\dfrac{5}{8}$

3. $7\dfrac{11}{16}$
$+2\dfrac{3}{16}$

4. $\dfrac{9}{15}$
$-\dfrac{3}{15}$

5. 7
$-\dfrac{15}{24}$

6. $7\dfrac{1}{6}$
$-1\dfrac{5}{6}$

Helpful Hints	Use what you have learned to solve the following problems.

Find the least common denominator of each of the following.

S1. $\dfrac{7}{8}$ and $\dfrac{11}{12}$　　　　　S2. $\dfrac{2}{3}$, $\dfrac{5}{12}$ and $\dfrac{5}{6}$　　　　　1. $\dfrac{7}{9}$ and $\dfrac{5}{6}$

2. $\dfrac{3}{4}$ and $\dfrac{1}{7}$　　　　　3. $\dfrac{1}{2}$ and $\dfrac{7}{15}$　　　　　4. $\dfrac{7}{30}$ and $\dfrac{1}{45}$

5. $\dfrac{3}{4}$ and $\dfrac{7}{18}$　　　　　6. $\dfrac{1}{9}$ and $\dfrac{1}{12}$　　　　　7. $\dfrac{1}{4}$, $\dfrac{1}{2}$ and $\dfrac{1}{3}$

8. $\dfrac{1}{12}$ and $\dfrac{7}{16}$　　　　　9. $\dfrac{1}{9}$ and $\dfrac{11}{72}$　　　　　10. $\dfrac{1}{2}$, $\dfrac{1}{5}$ and $\dfrac{1}{6}$

1.
2.
3.
4.
5.
6.
7.
8.
9.
10.
Score

Problem Solving	Nina bought 12 gallons of paint to paint her house. If she used $8\dfrac{3}{4}$ gallons how much paint does she have left?

Review Exercises

1. $\dfrac{7}{12}$
    $+\dfrac{9}{12}$

2. $\dfrac{15}{16}$
    $-\dfrac{11}{16}$

3. Find the least common denominator for $\dfrac{5}{6}$ and $\dfrac{7}{15}$.

6. $\dfrac{5}{2}$
    $\dfrac{3}{2}$
    $+\dfrac{7}{2}$

4. Reduce $\dfrac{56}{70}$ to its lowest terms.

5. Change $\dfrac{57}{11}$ to a mixed number.

Helpful Hints	To add fractions with unlike denominators, find the least common denominator. Multiply each fraction by one to make equivalent fractions. Finally, add.	**Examples:**

$$\dfrac{2}{5} \times \dfrac{2}{2} = \dfrac{4}{10}$$
$$+\dfrac{1}{2} \times \dfrac{5}{5} = \dfrac{5}{10}$$
$$\dfrac{9}{10}$$

$$\dfrac{5}{6} \times \dfrac{2}{2} = \dfrac{10}{12}$$
$$+\dfrac{1}{4} \times \dfrac{3}{3} = \dfrac{3}{12}$$
$$\dfrac{13}{12} = 1\dfrac{1}{12}$$

S1. $\dfrac{1}{3}$
     $+\dfrac{1}{4}$

S2. $\dfrac{3}{5}$
     $+\dfrac{7}{10}$

1. $\dfrac{5}{9}$
    $+\dfrac{1}{3}$

2. $\dfrac{2}{3}$
    $+\dfrac{1}{2}$

3. $\dfrac{1}{4}$
    $+\dfrac{2}{3}$

4. $\dfrac{3}{4}$
    $+\dfrac{2}{3}$

5. $\dfrac{5}{6}$
    $+\dfrac{5}{12}$

6. $\dfrac{1}{2}$
    $+\dfrac{3}{4}$

7. $\dfrac{1}{6}$
    $+\dfrac{3}{4}$

8. $\dfrac{7}{9}$
    $+\dfrac{1}{4}$

9. $\dfrac{7}{11}$
    $+\dfrac{1}{2}$

10. $\dfrac{3}{8}$
     $+\dfrac{1}{6}$

1.

2.

3.

4.

5.

6.

7.

8.

9.

10.

Problem Solving

Frankie worked for $7\dfrac{1}{4}$ hours on Tuesday and $5\dfrac{3}{4}$ on Wednesday. How many more hours did he work on Tuesday then on Wednesday?

Score

Review Exercises

1. $\dfrac{9}{10}$
    $+\dfrac{3}{10}$

2. $3\dfrac{3}{8}$
    $+4\dfrac{7}{8}$

3. 5
    $-\dfrac{14}{20}$

4. 6
    $-3\dfrac{3}{8}$

5. $7\dfrac{1}{4}$
    $-2\dfrac{3}{4}$

6. Find the least common denominator for $\dfrac{5}{8}$, $\dfrac{1}{3}$ and $\dfrac{7}{12}$.

Helpful Hints	Use what you have learned to solve the following problems. *Remember, Change all improper fractions to mixed numerals. Reduce all fractions to lowest terms.

S1. $\dfrac{2}{3}$
     $+\dfrac{1}{9}$

S2. $\dfrac{3}{4}$
     $+\dfrac{5}{6}$

1. $\dfrac{7}{12}$
    $+\dfrac{1}{3}$

2. $\dfrac{1}{2}$
    $\dfrac{1}{4}$
    $+\dfrac{2}{5}$

3. $\dfrac{7}{25}$
    $+\dfrac{3}{5}$

4. $\dfrac{11}{12}$
    $+\dfrac{1}{2}$

5. $\dfrac{11}{15}$
    $+\dfrac{1}{3}$

6. $\dfrac{2}{3}$
    $+\dfrac{3}{5}$

7. $\dfrac{2}{3}$
    $\dfrac{1}{2}$
    $+\dfrac{3}{8}$

8. $\dfrac{3}{11}$
    $+\dfrac{5}{33}$

9. $\dfrac{5}{12}$
    $+\dfrac{3}{8}$

10. $\dfrac{4}{6}$
     $+\dfrac{1}{10}$

1.

2.

3.

4.

5.

6.

7.

8.

9.

10.

Score

Problem Solving	Kareem needs 70 dollars for a new cell phone. On Monday he earned $17\dfrac{1}{2}$ dollars and on Wednesday he earned $16\dfrac{1}{2}$ dollars. How much more does he need to have enough to buy the cell phone?

103

Review Exercises

1. $\dfrac{2}{5}$
    $+\dfrac{1}{5}$

2. $\dfrac{5}{6}$
    $+\dfrac{1}{6}$

3. $3\dfrac{4}{5}$
    $+6\dfrac{3}{5}$

4. $\dfrac{2}{3}$
    $+\dfrac{1}{4}$

5. $\dfrac{3}{4}$
    $+\dfrac{2}{5}$

6. $\dfrac{7}{12}$
    $+\dfrac{2}{3}$

| **Helpful Hints** | To subtract fractions with unlike denominators, find the least common denominator. Multiply each fraction by one to make equivalent fractions. Finally, subtract. Reduce answers to lowest terms. | **Examples:** $\dfrac{3}{5} \times \dfrac{2}{2} = \dfrac{6}{10}$ $\dfrac{5}{6} \times \dfrac{2}{2} = \dfrac{10}{12}$
 $-\dfrac{1}{2} \times \dfrac{5}{5} = \dfrac{5}{10}$ $-\dfrac{1}{4} \times \dfrac{3}{3} = \dfrac{3}{12}$
 $\dfrac{1}{10}$ $\dfrac{7}{12}$ |

S1. $\dfrac{5}{9}$
     $-\dfrac{1}{3}$

S2. $\dfrac{5}{6}$
     $-\dfrac{1}{4}$

1. $\dfrac{7}{8}$
    $-\dfrac{4}{5}$

2. $\dfrac{9}{10}$
    $-\dfrac{1}{3}$

3. $\dfrac{4}{5}$
    $-\dfrac{1}{6}$

4. $\dfrac{11}{12}$
    $-\dfrac{2}{3}$

5. $\dfrac{5}{6}$
    $-\dfrac{2}{3}$

6. $\dfrac{11}{18}$
    $-\dfrac{2}{9}$

7. $\dfrac{5}{6}$
    $-\dfrac{7}{12}$

8. $\dfrac{4}{5}$
    $-\dfrac{1}{2}$

9. $\dfrac{7}{8}$
    $-\dfrac{2}{7}$

10. $\dfrac{11}{15}$
     $-\dfrac{1}{3}$

1.
2.
3.
4.
5.
6.
7.
8.
9.
10.

Score

Problem Solving

Jamie weighed $110\dfrac{1}{4}$ pounds two months ago. If she now weighs 115 pounds, how many pounds did she gain?

Review Exercises

1. $\dfrac{4}{5}$
$-\dfrac{1}{5}$

2. $\dfrac{7}{8}$
$-\dfrac{1}{8}$

3. $3\dfrac{11}{12}$
$-2\dfrac{3}{12}$

4. 7
$-2\dfrac{11}{12}$

5. 9
$-\dfrac{3}{4}$

6. $\dfrac{5}{6}$
$-\dfrac{1}{5}$

Helpful Hints	Use what you have learned to solve the following problems. *Remember, Reduce your answers to lowest terms.

S1. $\dfrac{3}{4}$
$-\dfrac{1}{2}$

S2. $\dfrac{7}{8}$
$-\dfrac{1}{5}$

1. $\dfrac{5}{6}$
$-\dfrac{1}{3}$

2. $\dfrac{3}{4}$
$-\dfrac{3}{16}$

3. $\dfrac{5}{8}$
$-\dfrac{1}{4}$

4. $\dfrac{4}{5}$
$-\dfrac{1}{20}$

5. $\dfrac{3}{4}$
$-\dfrac{1}{3}$

6. $\dfrac{7}{9}$
$-\dfrac{1}{18}$

7. $\dfrac{4}{5}$
$-\dfrac{1}{2}$

8. $\dfrac{11}{15}$
$-\dfrac{1}{3}$

9. $\dfrac{1}{4}$
$-\dfrac{1}{6}$

10. $\dfrac{7}{8}$
$-\dfrac{1}{6}$

1.

2.

3.

4.

5.

6.

7.

8.

9.

10.

Score

Problem Solving	Sally wants to send holiday cards to 75 people. If cards come in boxes of 12, how many boxes does she need to buy? How many cards will be left over?

Review Exercises

1. $\dfrac{7}{8}$
   $+\ \dfrac{7}{8}$

2. $\dfrac{4}{5}$
   $+\ \dfrac{4}{15}$

3. $\dfrac{2}{3}$
   $+\ \dfrac{1}{5}$

4. $\dfrac{3}{8}$
   $-\ \dfrac{1}{8}$

5. 7
   $-\ 2\,\dfrac{3}{5}$

6. $\dfrac{4}{5}$
   $-\ \dfrac{2}{3}$

Helpful Hints	Use what you have learned to solve the following problems. *Remember* 1. Change improper fractions to mixed numerals. 2. Reduce all answers to lowest terms.

S1. $\dfrac{7}{8}$
    $-\ \dfrac{1}{4}$

S2. $\dfrac{5}{8}$
    $+\ \dfrac{3}{4}$

1. $\dfrac{5}{9}$
   $+\ \dfrac{1}{3}$

2. $\dfrac{3}{4}$
   $-\ \dfrac{2}{3}$

3. $\dfrac{14}{15}$
   $-\ \dfrac{1}{3}$

4. $\dfrac{4}{5}$
   $+\ \dfrac{1}{2}$

5. $\dfrac{11}{22}$
   $+\ \dfrac{1}{11}$

6. $\dfrac{4}{5}$
   $-\ \dfrac{3}{10}$

7. $\dfrac{3}{5}$
   $-\ \dfrac{1}{10}$

8. $\dfrac{11}{25}$
   $+\ \dfrac{2}{5}$

9. $\dfrac{4}{7}$
   $-\ \dfrac{3}{14}$

10. $\dfrac{11}{20}$
    $+\ \dfrac{3}{10}$

1.

2.

3.

4.

5.

6.

7.

8.

9.

10.

Score

Problem Solving	Rhonda bought $\dfrac{3}{4}$ pounds of white chocolate candy and $\dfrac{2}{3}$ pounds of dark chocolate candy. How many pounds of candy did she by altogether?

Review Exercises

1. $3\dfrac{3}{8}$
$+4\dfrac{1}{8}$

2. $5\dfrac{4}{5}$
$+6\dfrac{3}{5}$

3. $5\dfrac{7}{10}$
$+4\dfrac{9}{10}$

4. $6\dfrac{3}{4}$
$-2\dfrac{1}{4}$

5. $7\dfrac{1}{10}$
$-3\dfrac{7}{10}$

6. 9
$-\dfrac{12}{15}$

Helpful Hints	Use what you have learned to solve the following problems. * Be careful to find the **lowest** common denominator. * Reduce answers to **lowest** terms.

S1. $\dfrac{7}{9}$
$-\dfrac{5}{18}$

S2. $\dfrac{5}{6}$
$+\dfrac{1}{4}$

1. $\dfrac{11}{12}$
$-\dfrac{1}{6}$

2. $\dfrac{11}{20}$
$+\dfrac{1}{5}$

3. $\dfrac{2}{15}$
$+\dfrac{3}{10}$

4. $\dfrac{9}{16}$
$-\dfrac{1}{2}$

5. $\dfrac{7}{8}$
$-\dfrac{1}{6}$

6. $\dfrac{2}{5}$
$+\dfrac{1}{3}$

7. $\dfrac{5}{8}$
$-\dfrac{1}{2}$

8. $\dfrac{3}{8}$
$+\dfrac{1}{2}$

9. $\dfrac{21}{25}$
$-\dfrac{7}{50}$

10. $\dfrac{8}{9}$
$+\dfrac{1}{3}$

1.

2.

3.

4.

5.

6.

7.

8.

9.

10.

Score

Problem Solving	Ricardo worked 8 hours on Monday. He spent $3\dfrac{3}{4}$ hours working on the computer. How much time did he spend working on other activities?

Review Exercises

1. Change $3\frac{2}{7}$ to an improper fraction.

2. Change $\frac{17}{15}$ to a mixed numeral.

3. Reduce $\frac{16}{28}$ to its lowest terms.

4. Shade the figure showing $\frac{3}{4}$.

5. $\frac{4}{8}$
   $+ \frac{5}{8}$

6. $\frac{3}{4}$
   $+ \frac{3}{8}$

Helpful Hints	When adding mixed numerals with unlike denominators, first add the fractions. If there is an improper fraction, make it a mixed numeral. Finally, add the sum to the sum of the whole numbers. * Reduce fractions to lowest terms.	**Example:** $3\frac{2}{3} \times \frac{2}{2} = \frac{4}{6}$ $+ \; 2\frac{1}{2} \times \frac{3}{3} = \frac{3}{6}$ $\overline{5 \qquad\qquad \frac{7}{6} = 1\frac{1}{6} = \boxed{6\frac{1}{6}}}$

S1. $3\frac{1}{3}$
   $+4\frac{2}{5}$

S2. $5\frac{3}{4}$
   $+2\frac{1}{3}$

1. $3\frac{3}{4}$
   $+1\frac{1}{2}$

2. $5\frac{2}{3}$
   $+2\frac{1}{6}$

3. $4\frac{1}{5}$
   $+3\frac{7}{10}$

4. $3\frac{7}{10}$
   $+4\frac{1}{2}$

5. $7\frac{2}{3}$
   $+3\frac{5}{9}$

6. $2\frac{3}{5}$
   $+4\frac{1}{2}$

7. $5\frac{3}{7}$
   $+2\frac{11}{14}$

8. $6\frac{11}{18}$
   $+4\frac{4}{9}$

9. $5\frac{1}{3}$
   $+2\frac{1}{7}$

10. $7\frac{5}{6}$
   $+3\frac{3}{8}$

1.
2.
3.
4.
5.
6.
7.
8.
9.
10.

Score

Problem Solving

It rained $7\frac{3}{4}$ inches in January. In February it rained $9\frac{1}{2}$ inches. What was the total rainfall for both months?

Review Exercises

1. $\dfrac{3}{4}$
    $+ \dfrac{1}{3}$

2. $\dfrac{5}{7}$
    $+ \dfrac{1}{3}$

3. $\dfrac{7}{16}$
    $+ \dfrac{7}{8}$

4. $\dfrac{9}{10}$
    $- \dfrac{3}{5}$

5. $\dfrac{11}{15}$
    $- \dfrac{7}{30}$

6. $\dfrac{5}{8}$
    $- \dfrac{1}{8}$

Helpful Hints	Use what you have learned to solve the following problems. * Change improper fractions to mixed numerals. * Reduce all answers to **lowest** terms.

S1. $5\dfrac{1}{2}$
     $+3\dfrac{3}{4}$

S2. $4\dfrac{11}{15}$
     $+3\dfrac{13}{30}$

1. $7\dfrac{9}{10}$
    $+3\dfrac{7}{20}$

2. $6\dfrac{4}{5}$
    $+2\dfrac{1}{10}$

3. $7\dfrac{5}{6}$
    $+2\dfrac{1}{3}$

4. $9\dfrac{4}{7}$
    $+3\dfrac{5}{14}$

5. $2\dfrac{3}{4}$
    $+2\dfrac{3}{8}$

6. $7\dfrac{1}{2}$
    $+8\dfrac{2}{3}$

7. $4\dfrac{3}{4}$
    $+2\dfrac{2}{5}$

8. $6\dfrac{4}{5}$
    $+7\dfrac{3}{10}$

9. $5\dfrac{2}{3}$
    $+ \dfrac{1}{9}$

10. $4\dfrac{5}{8}$
     $+3\dfrac{13}{32}$

1.
2.
3.
4.
5.
6.
7.
8.
9.
10.

Score

Problem Solving	Manuel had 20 dollars. He spent $7\dfrac{1}{2}$ dollars on school supplies and $3\dfrac{1}{4}$ dollars to watch a movie. How much money did he have left?

Review Exercises

1. $\frac{7}{8}$ $-\frac{1}{8}$

2. 7 $-2\frac{7}{16}$

3. $5\frac{3}{4}$ $-2\frac{1}{4}$

4. $6\frac{1}{3}$ $-2\frac{2}{3}$

5. $9\frac{1}{9}$ $-6\frac{7}{9}$

6. $\frac{5}{6}$ $-\frac{1}{3}$

Helpful Hints	To subtract mixed numerals with unlike denominators, first subtract the fractions. If the fractions cannot be subtracted, take one from the whole number, increase the fraction, then subtract.

Examples:

$$\begin{array}{r} 5\!\!\!\!\diagup6\,\frac{1}{6} = \frac{2}{12} + \frac{12}{12} = \frac{14}{12} \\ -\ 3\,\frac{1}{4} = \frac{3}{12} \\ \hline 2\,\frac{11}{12} \end{array} \qquad \begin{array}{r} 7\,\frac{1}{2} \times \frac{3}{3} = \frac{3}{6} \\ -\ 2\,\frac{1}{3} \times \frac{2}{2} = \frac{2}{6} \\ \hline 5 \qquad \frac{1}{6} = 5\,\frac{1}{6} \end{array}$$

S1. $5\frac{1}{2}$ $-2\frac{1}{3}$

S2. $6\frac{1}{3}$ $-2\frac{1}{2}$

1. $7\frac{3}{4}$ $-2\frac{1}{2}$

2. $9\frac{4}{5}$ $-3\frac{1}{10}$

3. $6\frac{1}{5}$ $-2\frac{2}{3}$

4. $5\frac{2}{5}$ $-3\frac{7}{10}$

5. $6\frac{1}{7}$ $-2\frac{5}{14}$

6. $9\frac{1}{5}$ $-3\frac{9}{10}$

7. $5\frac{1}{3}$ $-2\frac{8}{9}$

8. $6\frac{7}{8}$ $-3\frac{5}{16}$

9. $3\frac{1}{6}$ $-1\frac{1}{4}$

10. $5\frac{1}{5}$ $-2\frac{7}{15}$

1.

2.

3.

4.

5.

6.

7.

8.

9.

10.

Score

Problem Solving	Paula needs 30 dollars for a new dress. She earned $8\frac{1}{2}$ dollars last week and another $5\frac{1}{4}$ dollars this week. How much more does she need in order to buy the dress?

Review Exercises

1. $3\frac{2}{5}$ 2. $5\frac{3}{4}$ 3. $6\frac{1}{8}$ 4. $7\frac{1}{3}$ 5. $8\frac{15}{16}$ 6. $\frac{1}{2}$

$+2\frac{1}{5}$ $+7\frac{3}{4}$ $+3\frac{5}{8}$ $+4\frac{3}{5}$ $+2\frac{1}{8}$ $\frac{1}{3}$

                                                                    $+\frac{1}{5}$

Helpful Hints	Use what you have learned to solve the following problems. * Take one from the whole number when necessary. * Reduce all answers to lowest terms.

S1. $6\frac{1}{4}$ S2. $5\frac{1}{4}$ 1. $6\frac{1}{2}$ 2. $7\frac{1}{3}$

   $-2\frac{4}{8}$ $-2\frac{9}{20}$ $-2\frac{1}{3}$ $-2\frac{1}{2}$

3. $5\frac{2}{9}$ 4. $4\frac{1}{6}$ 5. $7\frac{7}{8}$ 6. $5\frac{3}{10}$

   $-2\frac{1}{3}$ $-1\frac{1}{3}$ $-2\frac{7}{16}$ $-1\frac{17}{30}$

7. $11\frac{1}{5}$ 8. $5\frac{3}{10}$ 9. $8\frac{1}{2}$ 10. $9\frac{1}{8}$

   $-2\frac{9}{10}$ $-2\frac{7}{40}$ $-3\frac{3}{5}$ $-2\frac{2}{3}$

1.
2.
3.
4.
5.
6.
7.
8.
9.
10.

Score

Problem Solving	Victoria started with 20 dollars. She spent $7\frac{1}{4}$ dollars, then she earned $9\frac{1}{2}$ dollars. How much money does she have now?

Review Exercises

1. $\dfrac{1}{2}$

 $+ \dfrac{1}{3}$

2. $\dfrac{3}{4}$

 $- \dfrac{1}{8}$

3. $7 \dfrac{1}{8}$

 $- 4 \dfrac{3}{8}$

4. $7 \dfrac{2}{3}$

 $- 3 \dfrac{2}{3}$

5. $9 \dfrac{1}{2}$

 $- 2 \dfrac{2}{3}$

6. $7 \dfrac{3}{8}$

 $+ 3 \dfrac{3}{4}$

Helpful Hints	Use what you have learned to solve the following problems. *Be sure all fractions are reduced to lowest terms.

S1. $3 \dfrac{1}{5}$

 $- 2 \dfrac{2}{5}$

S2. $5 \dfrac{1}{3}$

 $- 2 \dfrac{5}{6}$

1. $\dfrac{7}{8}$

 $- \dfrac{1}{3}$

2. $\dfrac{8}{9}$

 $- \dfrac{1}{6}$

3. 6

 $- 2 \dfrac{3}{7}$

4. $6 \dfrac{1}{8}$

 $+ 7 \dfrac{3}{4}$

5. $9 \dfrac{7}{10}$

 $+ 8 \dfrac{9}{10}$

6. $7 \dfrac{1}{3}$

 $- 2 \dfrac{5}{9}$

7. $5 \dfrac{3}{4}$

 $- 2 \dfrac{2}{5}$

8. $9 \dfrac{1}{2}$

 $- 6$

9. $\dfrac{8}{15}$

 $+ \dfrac{7}{30}$

10. $6 \dfrac{2}{3}$

 $+ 3 \dfrac{5}{6}$

1.

2.

3.

4.

5.

6.

7.

8.

9.

10.

Score

Problem Solving	Juanito worked $7 \dfrac{1}{2}$ hours on Monday and $9 \dfrac{1}{2}$ hours on Tuesday. If she was paid 12 dollars per hour, how much did she earn?

Review Exercises

1. Write 2 fractions for the shaded part.

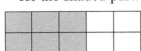

2. Change $7\frac{1}{2}$ to an improper fraction.

3. Find the missing numerator. $\frac{7}{8} = \frac{}{24}$

4. Change $\frac{31}{5}$ to a mixed numeral.

5. Reduce $\frac{40}{50}$ to its lowest terms.

6.
$$\begin{array}{r} \frac{3}{5} \\ \frac{4}{5} \\ +\ \frac{2}{5} \\ \hline \end{array}$$

Helpful Hints	Use what you have learned to solve the following problems. * If a particular problem gives you difficulty, it can be helpful to refer to the "Helpful Hints" sections from previous lessons.

S1.
$$\begin{array}{r} 9\frac{1}{2} \\ -\ 7 \\ \hline \end{array}$$

S2.
$$\begin{array}{r} 7 \\ -\ 1\frac{3}{4} \\ \hline \end{array}$$

1.
$$\begin{array}{r} \frac{3}{4} \\ +\ \frac{1}{8} \\ \hline \end{array}$$

2.
$$\begin{array}{r} \frac{9}{16} \\ +\ \frac{11}{16} \\ \hline \end{array}$$

3.
$$\begin{array}{r} \frac{7}{8} \\ -\ \frac{1}{6} \\ \hline \end{array}$$

4.
$$\begin{array}{r} 6\frac{1}{4} \\ -\ 3\frac{3}{4} \\ \hline \end{array}$$

5.
$$\begin{array}{r} 5\frac{7}{8} \\ +\ 6\frac{1}{8} \\ \hline \end{array}$$

6.
$$\begin{array}{r} 9\frac{5}{12} \\ -\ 2\frac{5}{6} \\ \hline \end{array}$$

7.
$$\begin{array}{r} 8\frac{4}{7} \\ +\ 3\frac{9}{14} \\ \hline \end{array}$$

8.
$$\begin{array}{r} \frac{2}{3} \\ \frac{1}{4} \\ +\ \frac{1}{5} \\ \hline \end{array}$$

9.
$$\begin{array}{r} \frac{11}{12} \\ +\ \frac{5}{24} \\ \hline \end{array}$$

10.
$$\begin{array}{r} 5\frac{4}{5} \\ +\ 2\frac{1}{2} \\ \hline \end{array}$$

1.

2.

3.

4.

5.

6.

7.

8.

9.

10.

Score

Problem Solving	It takes a man $15\frac{1}{4}$ to drive to work and $18\frac{1}{2}$ minutes to drive home. What is his total commute time?

Review Exercises

1. $3\frac{1}{2}$ $+2\frac{1}{3}$

2. $7\frac{1}{2}$ $-3\frac{2}{3}$

3. $\frac{9}{10}$ $+\frac{2}{5}$

4. 7 $-2\frac{15}{16}$

5. $9\frac{11}{12}$ $-2\frac{1}{12}$

6. $\frac{3}{4}$ $-\frac{1}{12}$

Helpful Hints

When multiplying common fractions, first multiply the numerators. Next, multiply the denominators. If the answer is an improper fraction, change it to a mixed numeral.

Examples: $\frac{3}{4} \times \frac{2}{7} = \frac{6}{28} = \frac{3}{14}$ * Be sure to reduce fractions to lowest terms.

$\frac{3}{2} \times \frac{7}{8} = \frac{21}{16} = 1\frac{5}{16}$

S1. $\frac{3}{4} \times \frac{5}{7} =$

S2. $\frac{5}{6} \times \frac{2}{3} =$

1. $\frac{5}{9} \times \frac{1}{7} =$

2. $\frac{2}{5} \times \frac{1}{2} =$

3. $\frac{5}{2} \times \frac{3}{4} =$

4. $\frac{5}{9} \times \frac{2}{3} =$

5. $\frac{6}{5} \times \frac{8}{9} =$

6. $\frac{4}{3} \times \frac{4}{5} =$

7. $\frac{5}{2} \times \frac{9}{10} =$

8. $\frac{8}{7} \times \frac{4}{5} =$

9. $\frac{1}{2} \times \frac{4}{5} =$

10. $\frac{9}{2} \times \frac{3}{7} =$

1.	
2.	
3.	
4.	
5.	
6.	
7.	
8.	
9.	
10.	
Score	

Problem Solving

Susan picked three boxes of apples. The weights were $12\frac{3}{4}$ pounds, $13\frac{1}{2}$ pounds, and $11\frac{1}{4}$ pounds. What was the total weight of the apples?

Review Exercises

1. Change $2\frac{3}{8}$ to an improper fraction.

3. Change $\frac{9}{2}$ to a mixed numeral.

6. $\frac{3}{5}$
   $\frac{4}{5}$
   $+\frac{2}{5}$

5. $3\frac{2}{5}$
   -2

2. $\frac{3}{4} \times \frac{2}{3} =$

4. $\frac{2}{3} + \frac{1}{5} =$

| Helpful Hints | Use what you have learned to solve the following problems. When multiplying fractions, "of" means to multiply. **Example:** Find $\frac{1}{2}$ of $\frac{3}{4}$ means: $\frac{1}{2} \times \frac{3}{4}$ | * Be sure to reduce fractions to lowest terms. |

S1. Find $\frac{3}{4}$ of $\frac{4}{5} =$	S2. $\frac{7}{2} \times \frac{3}{4} =$	1. Find $\frac{1}{3}$ of $\frac{8}{5} =$	1.
			2.
			3.
			4.
2. $\frac{5}{3} \times \frac{3}{4} =$	3. $\frac{9}{10} \times \frac{3}{2} =$	4. $\frac{1}{2}$ of $\frac{8}{3} =$	5.
			6.
5. $\frac{3}{4} \times \frac{5}{6} =$	6. $\frac{5}{4} \times \frac{3}{2} =$	7. $\frac{7}{3} \times \frac{3}{2} =$	7.
			8.
			9.
8. Find $\frac{2}{3}$ of $\frac{5}{6} =$	9. $\frac{5}{6} \times \frac{3}{2} =$	10. $\frac{3}{4} \times \frac{8}{9} =$	10.
			Score

| Problem Solving | Julia bought a CD that cost 23 dollars. If she paid for it with two twenty dollar bills, how much change should she receive back? |

Review Exercises

1. $\dfrac{1}{2}$ of $\dfrac{3}{4}$ =

3. $\dfrac{3}{5}$
   $+\ \dfrac{1}{2}$

4. $5\dfrac{1}{2}$
   $-2\dfrac{1}{3}$

5. $7\dfrac{3}{4}$
   -4

6. $9\dfrac{1}{3}$
   $-2\dfrac{2}{3}$

2. $\dfrac{5}{2} \times \dfrac{2}{3}$ =

Helpful Hints	If the numerator of one fraction and the denominator of another have a common factor, they can be divided out before you multiply the fractions.	**Examples:** $\dfrac{3}{\cancel{4}_{1}} \times \dfrac{\cancel{8}^{2}}{11} = \dfrac{6}{11}$ $\dfrac{7}{\cancel{8}_{4}} \times \dfrac{\cancel{6}^{3}}{5} = \dfrac{21}{20} = 1\dfrac{1}{20}$ 4 is a common factor. 2 is a common factor.

S1. $\dfrac{3}{5} \times \dfrac{7}{9}$ =

S2. $\dfrac{12}{15} \times \dfrac{5}{6}$ =

1. $\dfrac{3}{5} \times \dfrac{15}{16}$ =

2. $\dfrac{4}{15} \times \dfrac{5}{16}$ =

3. $\dfrac{5}{6}$ of $\dfrac{11}{15}$ =

4. $\dfrac{7}{3} \times \dfrac{11}{7}$ =

5. $\dfrac{5}{8} \times \dfrac{12}{25}$ =

6. $\dfrac{8}{9} \times \dfrac{3}{4}$ =

7. $\dfrac{3}{4} \times \dfrac{8}{15}$ =

8. $\dfrac{3}{4}$ of $\dfrac{3}{5}$ =

9. $\dfrac{5}{3} \times \dfrac{6}{7}$ =

10. $\dfrac{11}{6} \times \dfrac{4}{7}$ =

1.	
2.	
3.	
4.	
5.	
6.	
7.	
8.	
9.	
10.	
Score	

Problem Solving   There are 12 rows of seats in a theater. Each row has 15 seats. If 97 seats are occupied, how many seats are empty?

Review Exercises

1. Change $\dfrac{25}{15}$ to a mixed numeral.

2. $\dfrac{3}{4}$ of $\dfrac{5}{6} =$

3. $\dfrac{7}{2} \times \dfrac{1}{3} =$

4. Change $12\dfrac{1}{2}$ to an improper fraction.

5. $\begin{array}{r} 3\dfrac{5}{6} \\ +\,2\dfrac{1}{3} \\ \hline \end{array}$

6. $\begin{array}{r} \dfrac{11}{16} \\ -\,\dfrac{1}{4} \\ \hline \end{array}$

Helpful Hints	Use what you have learned to solve the following problems. * Change improper fractions to mixed numerals. * Reduce fractions to lowest terms.

S1. $\dfrac{5}{6} \times \dfrac{12}{25} =$

S2. Find $\dfrac{7}{8}$ of $\dfrac{24}{21} =$

1. $\dfrac{9}{10} \times \dfrac{7}{18} =$

2. $\dfrac{5}{6} \times \dfrac{8}{9} =$

3. $\dfrac{5}{6} \times \dfrac{11}{20} =$

4. $\dfrac{18}{25} \times \dfrac{50}{9} =$

5. $\dfrac{1}{2}$ of $\dfrac{9}{2} =$

6. $\dfrac{7}{6} \times \dfrac{3}{14} =$

7. $\dfrac{11}{3} \times \dfrac{9}{22} =$

8. $\dfrac{4}{3} \times \dfrac{15}{16} =$

9. $\dfrac{9}{10}$ of $\dfrac{40}{27} =$

10. $\dfrac{8}{5} \times \dfrac{11}{24} =$

1.

2.

3.

4.

5.

6.

7.

8.

9.

10.

Score

Problem Solving	The Gonzales family has a yearly income of 156,000 dollars. What is their average monthly income? (Hint: 1 year = 12 months).

Review Exercises

1. Reduce $\dfrac{40}{48}$ to lowest terms.

2. Change $\dfrac{27}{2}$ to a mixed numeral.

3. Change $5\dfrac{4}{5}$ to an improper fraction.

4. $\dfrac{3}{4} \times \dfrac{16}{17} =$

5. Find $\dfrac{2}{3}$ of $\dfrac{21}{40} =$

6. $\dfrac{7}{3} \times \dfrac{13}{14} =$

Helpful Hints	When multiplying whole numbers and fractions, write the whole number as a fraction and then multiply. **See examples:**	$\dfrac{2}{3} \times 15 =$ $\dfrac{2}{3} \times \overset{5}{\cancel{15}} = \dfrac{10}{1} = 10$	$\dfrac{3}{4} \times 9 =$ $\dfrac{3}{4} \times \dfrac{9}{1} = \dfrac{27}{4}$	$\begin{array}{r} 6\frac{3}{4} \\ 4\overline{)27} \\ \underline{24} \\ 3 \end{array}$

S1. $\dfrac{4}{5} \times 15 =$

S2. $\dfrac{3}{4} \times 5 =$

1. $\dfrac{3}{4} \times 16 =$

2. $12 \times \dfrac{5}{6} =$

3. $\dfrac{3}{4}$ of $24 =$

4. $\dfrac{3}{7} \times 5 =$

5. $\dfrac{1}{2} \times 37 =$

6. $\dfrac{1}{10} \times 15 =$

7. $6 \times \dfrac{7}{18} =$

8. Find $\dfrac{5}{6}$ of $9 =$

9. $\dfrac{7}{8} \times 40 =$

10. $\dfrac{3}{5} \times 7 =$

1.
2.
3.
4.
5.
6.
7.
8.
9.
10.
Score

Problem Solving	A class has 35 students. If $\dfrac{2}{5}$ of them are girls, how many girls are there in the class. (Hint: "of" means to multiply).

Review Exercises

1. $\frac{4}{5}$ $+ \frac{4}{5}$

2. $3\frac{1}{2}$ $+ 4\frac{5}{6}$

3. $7\frac{11}{12}$ $+ 2\frac{1}{2}$

4. $\frac{11}{15}$ $- \frac{1}{15}$

5. $\frac{11}{20}$ $- \frac{3}{10}$

6. $4\frac{1}{8}$ $- 3\frac{3}{4}$

Helpful Hints	Use what you have learned to solve the following problems. * Change improper fractions to mixed numerals. * Reduce fractions to lowest terms.

S1. $\frac{4}{5} \times 7 =$

S2. $\frac{5}{8} \times 40 =$

1. $\frac{4}{5}$ of $25 =$

2. $\frac{4}{5} \times 12 =$

3. $\frac{2}{3} \times 10 =$

4. $\frac{3}{4}$ of $60 =$

5. $\frac{1}{4} \times 15 =$

6. $\frac{2}{3} \times 30 =$

7. $55 \times \frac{7}{11} =$

8. $\frac{5}{6} \times 20 =$

9. $12 \times \frac{2}{5} =$

10. $20 \times \frac{1}{6} =$

1.

2.

3.

4.

5.

6.

7.

8.

9.

10.

Score

Problem Solving	A school has 300 students. $\frac{2}{5}$ of the students are boys. How many of the students are girls?

Review Exercises

1. $\dfrac{2}{3} \times \dfrac{5}{7} =$

2. $\dfrac{5}{6} \times \dfrac{12}{13} =$

3. $\dfrac{24}{25} \times \dfrac{50}{8} =$

4.   Find $\dfrac{2}{5}$ of $25 =$

5. $7 \times \dfrac{2}{3} =$

6. $\dfrac{3}{4} \times 32 =$

Helpful Hints

To multiply mixed numerals, first change them to improper fractions, then multiply. Express answers in lowest terms.

Example: $1\dfrac{1}{2} \times 1\dfrac{5}{6} =$

$\dfrac{\cancel{3}^{1}}{2} \times \dfrac{11}{\cancel{6}_{2}} = \dfrac{11}{4} = 2\dfrac{3}{4}$

S1. $\dfrac{1}{3} \times 1\dfrac{1}{3} =$ S2. $2\dfrac{1}{4} \times 2\dfrac{1}{3} =$ 1. $\dfrac{1}{3} \times 3\dfrac{1}{2} =$

2. $3\dfrac{1}{3} \times 2\dfrac{1}{5} =$ 3. $4 \times 2\dfrac{3}{4} =$ 4. $3\dfrac{1}{7} \times 1\dfrac{2}{5} =$

5. $3\dfrac{2}{3} \times 2\dfrac{1}{4} =$ 6. $2\dfrac{1}{2} \times 3\dfrac{1}{4} =$ 7. $2\dfrac{1}{2} \times 3\dfrac{1}{2} =$

8. $2\dfrac{1}{2} \times 8 =$ 9. $3\dfrac{1}{2} \times 4\dfrac{2}{3} =$ 10. $3\dfrac{1}{6} \times \dfrac{6}{7} =$

1.

2.

3.

4.

5.

6.

7.

8.

9.

10.

Score

Problem Solving

If a long distance runner can run 8 miles in an hour, how far can he run in $4\dfrac{1}{2}$ hours?

Review Exercises

1. $\dfrac{7}{8} \times \dfrac{4}{9} =$

2. $5 \times \dfrac{3}{4} =$

3. $\dfrac{2}{3} \times 24 =$

4. $7 \dfrac{1}{2}$
    $- 1 \dfrac{1}{3}$
    _____

5. $5 \dfrac{1}{3}$
    $- 2 \dfrac{2}{5}$
    _____

6. $3 \dfrac{2}{5}$
    $+ 4 \dfrac{7}{10}$
    _____

Helpful Hints	Use what you have learned to solve the following problems. * Divide out common factors. * Reduce fractions to lowest terms.

S1. $3 \dfrac{1}{2} \times 1 \dfrac{1}{7} =$

S2. $2 \dfrac{2}{3} \times 3 \dfrac{3}{4} =$

1. $\dfrac{9}{10} \times 2 \dfrac{1}{2} =$

2. $3 \dfrac{1}{8} \times 2 \dfrac{1}{5} =$

3. $3 \dfrac{1}{2} \times 2 \dfrac{1}{2} =$

4. $9 \times 2 \dfrac{2}{3} =$

5. $5 \dfrac{1}{3} \times 2 \dfrac{3}{4} =$

6. $4 \dfrac{1}{2} \times \dfrac{3}{4} =$

7. $5 \dfrac{1}{4} \times 2 \dfrac{1}{7} =$

8. $6 \dfrac{1}{2} \times 1 \dfrac{1}{3} =$

9. $3 \dfrac{1}{2} \times 2 \dfrac{1}{3} =$

10. $6 \dfrac{1}{4} \times 1 \dfrac{3}{5} =$

1.

2.

3.

4.

5.

6.

7.

8.

9.

10.

Score

Problem Solving	If a factory can produce 50 cars per day, how many cars can be produced in $5 \dfrac{1}{2}$ days?

Review Exercises

1. $\dfrac{2}{3} \times \dfrac{7}{8} =$　　　　2. $\dfrac{3}{4} \times 16 =$　　　　3. $\dfrac{2}{3} \times 2\dfrac{1}{2} =$

4. $5 \times 3\dfrac{1}{2} =$　　　　5. $2\dfrac{1}{2} \times 1\dfrac{1}{5} =$　　　　6. $3\dfrac{1}{3} \times 1\dfrac{1}{10} =$

Helpful Hints	Use what you have learned to solve the following problems. * Be sure to express answers in lowest terms. * Sometimes common factors may be divided out before you multiply.

S1. $\dfrac{7}{8} \times \dfrac{4}{9} =$　　　S2. $3\dfrac{1}{2} \times 1\dfrac{1}{7} =$　　　1. $\dfrac{4}{5} \times \dfrac{3}{8} =$

2. $\dfrac{9}{4} \times \dfrac{8}{11} =$　　　3. $\dfrac{3}{5} \times 45 =$　　　4. $\dfrac{3}{8} \times 5 =$

5. $\dfrac{3}{5} \times 3\dfrac{1}{2} =$　　　6. $2\dfrac{2}{3} \times \dfrac{1}{3} =$　　　7. $4 \times 2\dfrac{3}{4} =$

8. $1\dfrac{3}{5} \times 1\dfrac{1}{3} =$　　9. $4\dfrac{1}{6} \times 4\dfrac{4}{5} =$　　10. $9 \times 1\dfrac{2}{5} =$

1.

2.

3.

4.

5.

6.

7.

8.

9.

10.

Score

Problem Solving	Chuck has 240 dollars in his savings account. If he spends $\dfrac{3}{4}$ of his savings, how much does he have left in his account?

Review Exercises

1. $\frac{3}{4}$
    $-\frac{1}{2}$

2. $\frac{7}{8}$
    $-\frac{1}{3}$

3. 7
    $-2\frac{14}{15}$

4. $7\frac{1}{5}$
    $-2\frac{4}{5}$

5. $9\frac{5}{6}$
    $-1\frac{1}{4}$

6. $3\frac{2}{5}$
    $-1\frac{7}{10}$

Helpful Hints	Use what you have learned to solve the following problems. * If you have difficulty, refer to previous "Helpful Hints" sections for help.

S1. $\frac{3}{4}$ of $3\frac{1}{3} =$

S2. $2\frac{1}{4} \times 3\frac{1}{2} =$

1. $\frac{15}{16} \times \frac{8}{30} =$

2. $\frac{7}{2} \times \frac{9}{14} =$

3. $15 \times \frac{3}{5} =$

4. $\frac{1}{5} \times 4\frac{1}{2} =$

5. $3\frac{3}{4} \times 1\frac{4}{5} =$

6. $60 \times 2\frac{1}{2} =$

7. $3\frac{2}{3} \times 1\frac{5}{11} =$

8. $6 \times 3\frac{3}{4} =$

9. $2\frac{1}{4} \times 2\frac{1}{3} =$

10. $6 \times 3\frac{1}{2} =$

1.
2.
3.
4.
5.
6.
7.
8.
9.
10.
Score

Problem Solving	In a class of 36 students, $\frac{2}{3}$ of them received A's on a test. How many students received A's.

Review Exercises

1. $\frac{3}{4}$ of $1\frac{1}{5}$ =

2. $\frac{25}{21} \times \frac{7}{30}$ =

3. $5\frac{3}{8}$
    $+ 3\frac{3}{4}$
    _____

5. $3\frac{1}{5}$
    $- 1\frac{3}{5}$
    _____

4. $2\frac{1}{2} \times 5$ =

6. $2\frac{1}{2} \times 1\frac{4}{5}$ =

| **Helpful Hints** | To find the reciprocal of a common fraction, invert the fraction. To find the reciprocal of a mixed numeral, change the mixed numeral to an improper fraction, then invert it. To find the reciprocal of a whole number first make it a fraction then invert it. | **Examples:** The reciprocal of: $2\frac{1}{2}$ ($\frac{5}{2}$) is $\frac{2}{5}$ $\frac{3}{5}$ is $\frac{5}{3}$ or $1\frac{2}{3}$ $7(\frac{7}{1})$ is $\frac{1}{7}$ |

Find the reciprocal of each number.

S1. $\frac{2}{3}$

S2. $3\frac{1}{2}$

1. 7

2. $\frac{5}{8}$

3. $4\frac{1}{4}$

4. 15

5. $\frac{2}{7}$

6. $\frac{1}{9}$

7. 12

8. $\frac{2}{9}$

9. $5\frac{1}{2}$

10. 17

1.
2.
3.
4.
5.
6.
7.
8.
9.
10.

Problem Solving

Five students earned a total of 375 dollars. If they divided the money equally among themselves, how much did each student receive?

Score

Review Exercises

1. Change $7\frac{1}{2}$ to an improper fraction.

2. Reduce $\frac{18}{40}$ to its lowest terms.

3. Change $\frac{16}{7}$ to a mixed numeral.

4. Find the missing numerator.

$$\frac{7}{8} = \frac{}{56}$$

5. Shade in the fraction showing $\frac{3}{4}$.

6. Write 2 fractions for the shaded part.

Helpful Hints	Use what you have learned to solve the following problems

Find the reciprocal of each number.

S1. $\frac{7}{8}$

S2. $4\frac{3}{4}$

1. $\frac{11}{15}$

2. 6

3. $3\frac{1}{9}$

4. 16

5. $\frac{2}{15}$

6. $\frac{1}{6}$

7. 50

8. $\frac{3}{12}$

9. $7\frac{2}{3}$

10. $\frac{11}{16}$

1.
2.
3.
4.
5.
6.
7.
8.
9.
10.

Score

Problem Solving — If the normal temperature for an adult is $98\frac{3}{4}$ degrees, and a man has a temperature of 101 degrees, how much above normal is his temperature?

Review Exercises

1. $\dfrac{3}{4}$
    $+ \dfrac{2}{3}$

2. $\dfrac{3}{5}$ of $2\dfrac{1}{2} =$

3. $7 - 2\dfrac{1}{3} =$

4. $3\dfrac{1}{2} \times 1\dfrac{4}{7} =$

5. $\dfrac{2}{3}$
    $\dfrac{1}{2}$
    $+ \dfrac{1}{4}$

6. $3\dfrac{1}{8}$
    $- 1\dfrac{7}{8}$

| **Helpful Hints** | To divide fractions, find the reciprocal of the second number, then multiply the fractions. **Examples:** | $\dfrac{2}{3} \div \dfrac{1}{2} =$ $\dfrac{2}{3} \times \dfrac{2}{1} = \dfrac{4}{3} = \boxed{1\dfrac{1}{3}}$ | $2\dfrac{1}{2} \div 1\dfrac{1}{2} = \dfrac{5}{2} \div \dfrac{3}{2}$ $\dfrac{5}{\cancel{2}} \times \dfrac{\cancel{2}^{1}}{3} = \dfrac{5}{3} = \boxed{1\dfrac{2}{3}}$ |

S1. $\dfrac{3}{5} \div \dfrac{3}{8} =$

S2. $4\dfrac{1}{2} \div 2 =$

1. $\dfrac{5}{8} \div \dfrac{1}{6} =$

2. $\dfrac{1}{4} \div \dfrac{1}{3} =$

3. $7 \div \dfrac{2}{3} =$

4. $5\dfrac{1}{2} \div \dfrac{1}{2} =$

5. $1\dfrac{3}{4} \div \dfrac{3}{8} =$

6. $3\dfrac{1}{4} \div \dfrac{11}{12} =$

7. $1\dfrac{1}{3} \div 3 =$

8. $7\dfrac{1}{2} \div 2 =$

9. $7\dfrac{1}{2} \div 2\dfrac{1}{2} =$

10. $4\dfrac{2}{3} \div 2\dfrac{1}{2} =$

| 1. |
| 2. |
| 3. |
| 4. |
| 5. |
| 6. |
| 7. |
| 8. |
| 9. |
| 10. |
| Score |

Problem Solving

$4\dfrac{1}{2}$ yards if cloth is to be divided into pieces that are $\dfrac{1}{2}$ yards long. How many pieces will there be?

Review Exercises

1. $\dfrac{3}{4} \times \dfrac{7}{9} =$

2. $\dfrac{3}{4} \div \dfrac{1}{4} =$

3. $2\dfrac{1}{2} \times 2 =$

4. $2\dfrac{1}{2} \div 2 =$

5. $1\dfrac{1}{3} \times \dfrac{3}{4} =$

6. $1\dfrac{1}{3} \div \dfrac{3}{4} =$

Helpful Hints	Use what you have learned to solve the following problems. * Remember, find the reciprocal of the second number, then multiply.

S1. $\dfrac{7}{8} \div \dfrac{3}{4} =$ S2. $2\dfrac{1}{2} \div 1\dfrac{1}{4} =$ 1. $\dfrac{5}{6} \div \dfrac{1}{3} =$

2. $\dfrac{3}{7} \div \dfrac{1}{2} =$ 3. $5 \div 1\dfrac{1}{2} =$ 4. $6\dfrac{1}{2} \div \dfrac{1}{4} =$

5. $1\dfrac{5}{6} \div \dfrac{1}{2} =$ 6. $4\dfrac{1}{2} \div 1\dfrac{1}{3} =$ 7. $2\dfrac{2}{3} \div \dfrac{1}{3} =$

8. $3\dfrac{3}{4} \div 2 =$ 9. $4\dfrac{1}{2} \div 2\dfrac{2}{3} =$ 10. $12\dfrac{1}{2} \div 2\dfrac{1}{2} =$

1.
2.
3.
4.
5.
6.
7.
8.
9.
10.
Score

Problem Solving	A floor tile is $\dfrac{3}{4}$ inches thick. How many inches thick would a stack of 36 tiles be?

Review Exercises

1. $\dfrac{3}{8}$ 2. $\dfrac{7}{16}$ 3. $\dfrac{3}{4}$ of $1\dfrac{1}{2}$ =

   $+\ \dfrac{3}{4}$ $-\ \dfrac{1}{4}$

   _____   _____

4. $\dfrac{7}{8} \div \dfrac{1}{4} =$ 5. $3\dfrac{1}{2} \div \dfrac{3}{4} =$ 6. $2\dfrac{2}{3} \div 1\dfrac{1}{3} =$

Helpful Hints	Sometimes it is necessary to compare the values of fractions. **Example:** $\dfrac{3}{4}$ and $\dfrac{2}{3}$ LCD is 12.
	1. Find the least common denominator (LCD).
	2. Change each fraction to an equivalent fraction and compare. $\dfrac{3}{4} \times \dfrac{3}{3} = \boxed{\dfrac{9}{12}}$ $\dfrac{2}{3} \times \dfrac{4}{4} = \boxed{\dfrac{8}{12}}$ $\dfrac{3}{4}$ is the larger fraction.

Find the larger of each pair of fractions.

S1. $\dfrac{5}{8}$, $\dfrac{6}{7}$ S2. $\dfrac{7}{9}$, $\dfrac{5}{6}$ 1. $\dfrac{11}{12}$, $\dfrac{8}{9}$

2. $\dfrac{3}{5}$, $\dfrac{4}{7}$ 3. $\dfrac{7}{10}$, $\dfrac{5}{6}$ 4. $\dfrac{9}{11}$, $\dfrac{5}{8}$

5. $\dfrac{5}{6}$, $\dfrac{13}{15}$ 6. $\dfrac{2}{7}$, $\dfrac{3}{8}$ 7. $\dfrac{11}{15}$, $\dfrac{23}{30}$

8. $\dfrac{1}{6}$, $\dfrac{2}{11}$ 9. $\dfrac{9}{10}$, $\dfrac{8}{9}$ 10. $\dfrac{11}{12}$, $\dfrac{7}{8}$

1.
2.
3.
4.
5.
6.
7.
8.
9.
10.

Score

Problem Solving	A group of hikers need to travel 55 miles. They hike 6 miles per day. After 6 days, how much farther did they still have to hike?

Review Exercises

1. $\dfrac{3}{4} \div \dfrac{1}{2} =$ 2. $2\dfrac{1}{2} \div \dfrac{1}{4} =$ 3. $3\dfrac{3}{4} \div 1\dfrac{1}{4} =$

4. $\dfrac{3}{5}$ of $1\dfrac{1}{2} =$ 5. $5 \times 1\dfrac{3}{8} =$ 6. $7\dfrac{1}{2} \times 1\dfrac{2}{5} =$

Helpful Hints	Use what you have learned to solve the following problems.

Find the larger of each pair of fractions.

S1. $\dfrac{3}{4}$, $\dfrac{7}{9}$	S2. $\dfrac{11}{20}$, $\dfrac{2}{5}$	1. $\dfrac{11}{14}$, $\dfrac{3}{7}$
2. $\dfrac{10}{12}$, $\dfrac{4}{5}$	3. $\dfrac{3}{4}$, $\dfrac{4}{7}$	4. $\dfrac{1}{2}$, $\dfrac{5}{11}$
5. $\dfrac{2}{6}$, $\dfrac{3}{11}$	6. $\dfrac{11}{20}$, $\dfrac{3}{10}$	7. $\dfrac{5}{6}$, $\dfrac{3}{4}$
8. $\dfrac{5}{6}$, $\dfrac{7}{8}$	9. $\dfrac{5}{9}$, $\dfrac{3}{5}$	10. $\dfrac{2}{7}$, $\dfrac{3}{8}$

1.

2.

3.

4.

5.

6.

7.

8.

9.

10.

Score

Problem Solving	Ruth was born in 1989. How old will she be in 2012?

Fraction Concepts - Final Review

Reduce all answers to lowest terms.

1. Reduce $\dfrac{20}{25}$ to its lowest terms.

2. Reduce $\dfrac{24}{30}$ to its lowest terms.

3. Reduce $\dfrac{40}{48}$ to its lowest terms.

4. Change $\dfrac{11}{5}$ to a mixed numeral.

5. Change $\dfrac{28}{12}$ to a mixed numeral.

6. Change $\dfrac{87}{25}$ to a mixed numeral.

7. Change $5\dfrac{3}{5}$ to an improper fraction.

8. Change $12\dfrac{4}{5}$ to an improper fraction.

9. Change $7\dfrac{1}{3}$ to an improper fraction.

10. Find the missing numerator.
$$\dfrac{5}{6} = \dfrac{}{48}$$

11. Find the missing numerator.
$$\dfrac{}{49} = \dfrac{6}{7}$$

12. Find the missing numerator.
$$\dfrac{5}{6} = \dfrac{}{72}$$

13. Find the LCD. $\dfrac{1}{6}$, $\dfrac{1}{4}$, $\dfrac{1}{3}$

14. Find the LCD. $\dfrac{1}{5}$, $\dfrac{1}{3}$, $\dfrac{3}{4}$

15. Find the LCD. $\dfrac{5}{6}$, $\dfrac{1}{4}$, $\dfrac{1}{2}$

16. Write 2 fractions for the shaded part.

17. Write 2 fractions for the shaded part.

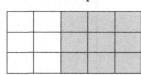

18. Shade in $\dfrac{3}{5}$ of the figure.

19. Find the larger fraction.
$$\dfrac{5}{8} , \dfrac{2}{3}$$

20. Find the larger fraction.
$$\dfrac{5}{6} , \dfrac{7}{9}$$

1.	
2.	
3.	
4.	
5.	
6.	
7.	
8.	
9.	
10.	
11.	
12.	
13.	
14.	
15.	
16.	
17.	
18.	
19.	
20.	

Fraction Operations - Final Review

Reduce all answers to lowest terms.

1. $\dfrac{3}{7}$
    $+ \dfrac{4}{7}$

2. $\dfrac{5}{8}$
    $+ \dfrac{7}{8}$

3. $\dfrac{3}{4}$
    $+ \dfrac{2}{5}$

4. $3\dfrac{3}{4}$
    $+2\dfrac{5}{8}$

5. $5\dfrac{2}{3}$
    $+3\dfrac{5}{6}$

6. $\dfrac{7}{8}$
    $- \dfrac{1}{8}$

7. $5\dfrac{1}{7}$
    $-2\dfrac{3}{7}$

8. 7
    $-2\dfrac{4}{7}$

9. $6\dfrac{3}{4}$
    $-2\dfrac{1}{2}$

10. $7\dfrac{1}{3}$
    $-2\dfrac{4}{5}$

11. $\dfrac{3}{4} \times \dfrac{2}{7} =$

12. $\dfrac{15}{16} \times \dfrac{7}{30} =$

13. $\dfrac{4}{5} \times 30 =$

14. $\dfrac{3}{4} \times 4\dfrac{2}{3} =$

15. $2\dfrac{1}{2} \times 3\dfrac{1}{2} =$

16. $\dfrac{5}{6} \div \dfrac{1}{3} =$

17. $3\dfrac{1}{2} \div \dfrac{1}{4} =$

18. $3\dfrac{3}{4} \div 2\dfrac{1}{2} =$

19. $3\dfrac{3}{4} \div 1\dfrac{3}{8} =$

20. $5 \div 2\dfrac{1}{5} =$

1.	
2.	
3.	
4.	
5.	
6.	
7.	
8.	
9.	
10.	
11.	
12.	
13.	
14.	
15.	
16.	
17.	
18.	
19.	
20.	

Fraction Concepts - Final Test

Reduce all answers to lowest terms.

1. Reduce $\dfrac{24}{28}$ to its lowest terms.

2. Reduce $\dfrac{18}{30}$ to its lowest terms.

3. Reduce $\dfrac{80}{100}$ to its lowest terms.

4. Change $\dfrac{13}{4}$ to a mixed numeral.

5. Change $\dfrac{19}{12}$ to a mixed numeral.

6. Change $\dfrac{20}{14}$ to a mixed numeral.

7. Change $2\dfrac{1}{8}$ to an improper fraction.

8. Change $8\dfrac{2}{3}$ to an improper fraction.

9. Change $2\dfrac{5}{6}$ to an improper fraction.

10. Find the missing numerator.

    $\dfrac{3}{12} = \dfrac{}{36}$

11. Find the missing numerator.

    $\dfrac{}{20} = \dfrac{4}{5}$

12. Find the missing numerator.

    $\dfrac{7}{8} = \dfrac{}{56}$

13. Find the LCD. $\dfrac{1}{3}$, $\dfrac{1}{2}$, $\dfrac{1}{5}$

14. Find the LCD. $\dfrac{3}{8}$, $\dfrac{5}{6}$

15. Find the LCD. $\dfrac{1}{2}$, $\dfrac{1}{5}$, $\dfrac{1}{6}$

16. Write 2 fractions for the shaded part.

    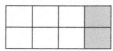

17. Write 2 fractions for the shaded part.

    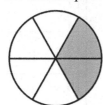

18. Shade in $\dfrac{3}{4}$ of the figure.

    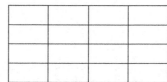

19. Find the larger fraction.

    $\dfrac{4}{7}$, $\dfrac{6}{11}$

20. Find the larger fraction.

    $\dfrac{3}{4}$, $\dfrac{4}{5}$, $\dfrac{1}{2}$

1.
2.
3.
4.
5.
6.
7.
8.
9.
10.
11.
12.
13.
14.
15.
16.
17.
18.
19.
20.

Fraction Operations - Final Test

Reduce all answers to lowest terms.

1. $\dfrac{3}{5}$
   $+ \dfrac{1}{5}$

2. $\dfrac{7}{12}$
   $+ \dfrac{11}{12}$

3. $\dfrac{3}{5}$
   $+ \dfrac{7}{10}$

4. $2\dfrac{3}{5}$
   $+ 3\dfrac{7}{15}$

5. $5\dfrac{3}{7}$
   $+ 2\dfrac{2}{3}$

6. $\dfrac{9}{10}$
   $- \dfrac{1}{10}$

7. $6\dfrac{1}{8}$
   $- 2\dfrac{3}{8}$

8. 6
   $- 2\dfrac{3}{8}$

9. $6\dfrac{3}{4}$
   $- 2\dfrac{1}{5}$

10. $7\dfrac{5}{12}$
    $- 2\dfrac{3}{4}$

11. $\dfrac{1}{2} \times \dfrac{8}{9} =$

12. $\dfrac{8}{15} \times \dfrac{5}{16} =$

13. $\dfrac{3}{4} \times 28 =$

14. $\dfrac{4}{5} \times 3\dfrac{1}{4} =$

15. $2\dfrac{2}{5} \times 2\dfrac{1}{2} =$

16. $\dfrac{3}{8} \div \dfrac{1}{4} =$

17. $3\dfrac{2}{3} \div \dfrac{1}{2} =$

18. $4\dfrac{1}{2} \div 2\dfrac{1}{3} =$

19. $5 \div 2\dfrac{1}{8} =$

20. $4\dfrac{1}{2} \div 3 =$

1.	
2.	
3.	
4.	
5.	
6.	
7.	
8.	
9.	
10.	
11.	
12.	
13.	
14.	
15.	
16.	
17.	
18.	
19.	
20.	

Section 3

Decimals and Percents

135

Review Exercises

Note to the students and teachers: This section will include daily review from all topics covered in this book. Here are some simple problems to get started.

1.   36 + 17 + 44 =

2.   521 - 306 =

3.   715
       - 264

4.   47
       348
     + 225

5.   Find the sum of 25, 36, and 48.

6.   Find the difference between 700 and 76.

Helpful Hints

$$9 . 8 \ 7 \ 6 \ 5 \ 4 \ 3$$

ones, tenths, hundredths, thousandths, ten-thousandths, hundred-thousandths, millionths

To read decimals, first read the whole number. Next, read the decimal point as "and." Next, read the number after the decimal point and its place value.

Examples:
    3.16 = three and sixteen hundredths
    14.011 = fourteen and eleven thousandths
    0.69 = sixty-nine hundredths

Write the following in words.

S1.   3.7   S2.   12.019

1.   0.87   2.   5.006

3.   115.7   4.   78.07

5.   6.3912   6.   0.085

7.   7.36   8.   9.002

9.   0.61   10.   2.333

1.
2.
3.
4.
5.
6.
7.
8.
9.
10.

Problem Solving

Juan has 709 dollars and Al has 529 dollars. How many more dollars does Juan have than Al?

Score

Review Exercises

1. 427
    816
     23
    + 142

2. Find the difference
    between 1,726 and 977.

3. 601 - 78 =

4. 37 + 42 + 36 =

5.   500
     - 276

6.   315
     x 4

Helpful Hints	Use what you have learned to solve the following problems.
	  * Read the decimal point as "and".
	  * Put hyphens in numbers between 20 and 99 when necessary.

Write the following in words.

S1. 3.006   S2. 0.0176

1. 0.8   2. 3.0005

3. 76.8   4. 7.008

5. 5.138   6. 0.015

7. 5.82   8. 4.03

9. 0.86   10. 4.224

1.

2.

3.

4.

5.

6.

7.

8.

9.

10.

Score

Problem Solving	Three classes at Hoover School have enrollments of 37, 48, and 40. What is the total enrollment of the three classes?

Review Exercises

1. 336
   19
   + 424

2. Write 2.007 in words.

3. 7,125
   - 743

4. Write 42.016 in words.

5. 37 + 16 + 274 =

6. Write 0.019 in words.

Helpful Hints

When reading, remember "and" means decimal point. The fraction part of a decimal ends in "th" or "ths." Be careful about placeholders.

Example:
Four and eight tenths = 4.8
Two hundred one and six hundredths = 201.06
One hundred four ten-thousandths = .0104

Write each of the following as a decimal. Use the chart at the bottom to help.

S1. Five and three hundredths.

S2. Four hundred thirty-six and eleven hundredths.

1. Seven and four tenths.

2. Twenty-two and fifteen thousandths.

3. Three hundred fifty-two ten-thousandths.

4. Seventy-four and forty-three thousandths.

5. Five hundred-thousandths.

6. Sixteen millionths.

7. Nine and forty-five thousandths.

8. Twenty and thirty-three ten-thousandths.

9. Eighty-six and nine tenths.

10. Eighty-six and nine millionths.

ones . tenths hundredths thousandths ten-thousandths hundred-thousandths millionths

9 . 8 7 6 5 4 3

1.	
2.	
3.	
4.	
5.	
6.	
7.	
8.	
9.	
10.	
Score	

Problem Solving

Crayons come in boxes of 24. How many crayons are there in fifteen boxes?

Review Exercises

1. Write 2.09 in words.

2. Write 2.009 in words.

3. Two and four hundredths equals what decimal?

4. 761 - 79 =

5. Fifteen thousandths equals what decimal?

6. 7,647
    362
    + 5,173

Helpful Hints

Use what you have learned to solve the following problems.
* "and" means decimal point.
* The fraction part of a decimal ends in "th" or "ths".
* Be careful about placeholders.

Write each of the following as a decimal. Use the chart at the bottom to help.

S1. Seven and sixty-two ten-thousandths..

S2. Two thousand nine hundred-thousandths.

1. Eight and nine hundredths.

2. Twelve and forty-one hundred-thousandths.

3. Forty-nine ten-thousandths.

4. Ninety-seven and five hundred thirteen millionths.

5. Forty-eight thousandths.

6. Fifty-two and eight tenths.

7. Five and four hundred ninety-six thousandths.

8. Three and five thousandths.

9. Twelve and thirty-three hundred-thousandths.

10. One hundred sixteen and five hundredths.

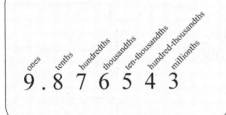

9 . 8 7 6 5 4 3
ones tenths hundredths thousandths ten-thousandths hundred-thousandths millionths

1.
2.
3.
4.
5.
6.
7.
8.
9.
10.
Score

Problem Solving

360 students are placed in fifteen equally-sized classes. How many students are in each class?

Review Exercises

1. Write 2.07 in words.

2. Write 7.017 in words.

3. Write seven and six thousandths as a decimal?

4. Write thirty-two millionths as a decimal?

5. Write 0.0017 in words.

6. Write five and eleven ten-thousandths as a decimal?

| **Helpful Hints** | When changing mixed numerals to decimals, remember to put a decimal after the whole number. | **Examples:** $3\frac{3}{10} = 3.3$ $\frac{16}{10,000} = .0016$ $42\frac{9}{10,000} = 42.0009$ $65\frac{12}{100,000} = 65.00012$ |

Write each of the following as a decimal. Use the chart at the bottom to help.

S1. $5\frac{6}{10}$

S2. $8\frac{9}{1,000}$

1. $21\frac{16}{100}$

2. $\frac{16}{100}$

3. $14\frac{17}{1,000}$

4. $119\frac{16}{100,000}$

5. $\frac{21}{10,000}$

6. $3\frac{196}{100,000}$

7. $4\frac{32}{1,000}$

8. $3\frac{324}{1,000}$

9. $4\frac{17}{1,000,000}$

10. $\frac{19}{10,000}$

9 . 8 7 6 5 4 3
ones tenths hundredths thousandths ten-thousandths hundred-thousandths millionths

1.	
2.	
3.	
4.	
5.	
6.	
7.	
8.	
9.	
10.	
Score	

Problem Solving

A plane traveled 3,150 miles in seven hours. What was its average speed per hour?

Review Exercises

1. Write ten and fourteen thousandths as a decimal.

2. Write 2.09 in words.

3. Write sixty-five thousandths as a decimal.

4. Write ten and fifteen hundred-thousandths as a decimal.

5. 724 - 617 =

6. Find the difference between 972 and 408.

Helpful Hints

Use what you have learned to solve the following problems.
* Remember to put a decimal point after the whole number.
* Be careful to use placeholders when necessary.

Write each of the following as a decimal. Use the chart at the bottom to help.

S1. $9\dfrac{17}{1,000}$

S2. $9\dfrac{17}{100,000}$

1. $42\dfrac{196}{1,000}$

2. $\dfrac{72}{1,000}$

3. $48\dfrac{8}{1,000}$

4. $16\dfrac{195}{100,000}$

5. $\dfrac{16}{1,000}$

6. $16\dfrac{119}{1,000,000}$

7. $4\dfrac{38}{10,000}$

8. $3\dfrac{176}{10,000}$

9. $\dfrac{71}{1,000}$

10. $6\dfrac{53}{10,000}$

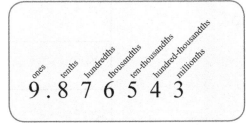

1.

2.

3.

4.

5.

6.

7.

8.

9.

10.

Score

Problem Solving

If a car can travel 32 miles for each gallon of gas that it uses, how far can the car travel using 12 gallons of gas?

Review Exercises

1. Write $9\frac{7}{100}$ as a decimal.

2. Write $1\frac{135}{100,000}$ as a decimal.

3. Write seventeen hundred-thousandths as a decimal.

4. Write 0.021 in words.

5. Write 3.19 in words.

6. Write six and thirteen ten-thousandths as a decimal.

| **Helpful Hints** | Decimals can easily be changed to mixed numbers and fractions. Remember that the whole number is to the left of the decimal. | **Examples:** | $2.6 = 2\frac{6}{10}$ $3.007 = 3\frac{7}{1,000}$ | $.210 = \frac{210}{1,000}$ $1.0019 = 1\frac{19}{10,000}$ |

Change the following to a mixed numeral or fraction. Use the chart at the bottom for help.

S1.   3.05

S2.   16.017

1.   45.0019

2.   .00005

3.   7.000016

4.   7.196

5.   79.6

6.   .07632

7.   14.00007

8.   16.024

9.   17.000145

10.   .00096

```
       ones  tenths  hundredths  thousandths  ten-thousandths  hundred-thousandths  millionths
        9  .   8       7            6             5                4                    3
```

1.
2.
3.
4.
5.
6.
7.
8.
9.
10.
Score

Problem Solving

A worker earned 750 dollars. He spent 300 dollars for his rent and 176 dollars for his car payment. How much of his earnings were left?

Review Exercises

1. Write eleven and six hundredths as a decimal.

2. Write six thousandths as a fraction.

3. Write $\frac{19}{1,000}$ as a decimal.

4. Write 7.006 as a mixed numeral.

5. Write six and three tenths as a mixed numeral.

6. Write seventy-two thousandths as a decimal.

Helpful Hints

Use what you have learned to solve the following problems.

* Remember, the whole number is to the left of the decimal.

Change the following to a mixed numeral or fraction.
Use the chart at the bottom for help.

S1. .0016	S2. 9.00125
1. 7.00009	2. .016
3. 7.29	4. 6.00002
5. 87.3	6. 5.0072
7. 15.000006	8. 42.1
	9. 163.0137
	10. .11234

1.
2.
3.
4.
5.
6.
7.
8.
9.
10.
Score

ones tenths hundredths thousandths ten-thousandths hundred-thousandths millionths

9 . 8 7 6 5 4 3

Problem Solving

June wants to send our invitations to 86 people. The invitations come in packages of 15. How many packages must she buy. How many cards will be left over?

Review Exercises

1. Write 7.0016 as a mixed numeral.

2. Write $7\frac{15}{1,000}$ as a decimal.

3. Write five and six tenths as a decimal.

4. Write 6.013 in words.

5. Write $3\frac{7}{1,000}$ in words.

6. Write 72.0012 as a mixed numeral.

Helpful Hints	Zeroes can be put to the right of a decimal without changing the value. This helps when comparing the value of decimals.	< means less than > means greater than	**Example:** Compare 4.3 and 4.28 4.3 = 4.30 so 4.3 > 4.28

Place > or < to compare each pair of decimals. Use the chart at the bottom for help.

S1. 7.46 ☐ 7.5

S2. .88 ☐ .879

1. 3.099 ☐ 3.1

2. 7.52 ☐ 7.396

3. 6.22 ☐ 6.31

4. 1.7 ☐ 1.69

5. 2.21 ☐ 1.99

6. .4 ☐ .71

7. 2.29 ☐ 2.4

8. 3.65 ☐ 3.589

9. 9.09 ☐ 9.2

10. 8.199 ☐ 8.2

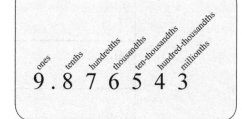

1.

2.

3.

4.

5.

6.

7.

8.

9.

10.

Score

Problem Solving

A car traveled 240 miles. For each 30 miles traveled, the car used one gallon of gas. How many gallons of gas were used on the trip? If gas cost 3 dollars per gallon, how much did the trip cost?

Review Exercises

1. Write .0129 as a fraction.

2. Write $7\frac{92}{1,000}$ in words.

3. Write .0135 in words.

4. Write seventeen and six thousandths as a decimal.

5. Change $15\frac{71}{10,000}$ to a decimal.

6. Change 3.096 to a mixed numeral.

Helpful Hints

Use what you have learned to solve the following problems.
* Add zeroes to help compare the values.

Place > or < to compare each pair of decimals. Use the chart at the bottom for help.

S1.   8.6 ☐ 8.72

S2.   .799 ☐ .81

1.   3.196 ☐ 3.2

2.   2.423 ☐ 2.398

3.   6.12 ☐ 6.115

4.   3.7 ☐ 3.68

5.   .9762 ☐ 1.001

6.   .4 ☐ .53

7.   2.41 ☐ 2.296

8.   4.63 ☐ 4.596

9.   .9 ☐ .875

10.   3.09 ☐ 3.112

9 . 8 7 6 5 4 3
ones tenths hundredths thousandths ten-thousandths hundred-thousandths millionths

1.
2.
3.
4.
5.
6.
7.
8.
9.
10.
Score

Problem Solving

Leo, Maria, Pedro, and Jill together earned 724 dollars. If they wanted to share the money equally, how much will each one receive?

Review Exercises

1. Change 52.623 to a mixed numeral.

2. Write $75 \frac{9}{1,000}$ as a decimal.

3.
$$\begin{array}{r} 624 \\ \times\ 7 \\ \hline \end{array}$$

4. Write fifty and nineteen thousandths as a decimal.

5. Write 6.23 in words.

6. 7,051 - 267 =

Helpful Hints	To add decimals, line up the decimal points and add as you would whole numbers. Write the decimal points in the answer. Zeroes may be placed to the right of the decimal.	**Example:** Add 3.16 + 2.4 + 12	$\begin{array}{r} 3.16 \\ 2.40 \\ +\ 12.00 \\ \hline 17.56 \end{array}$

S1.
$$\begin{array}{r} 3.16 \\ 12.4 \\ +\ 3.26 \\ \hline \end{array}$$

S2. 3.92 + 4.6 + .32 =

1. 32.16 + 1.7 + 7.493 =

2. 7.341 + 6.49 + .6 =

3.
$$\begin{array}{r} 7.64 \\ 19.633 \\ +\ 2.4 \\ \hline \end{array}$$

4. .37 + .6 + .73 =

5. 9.64 + 7 + 1.92 + .7 =

6. 72.163 + 11.4 + 63.42 =

7. .7 + .6 + .4 =

8. 17.33 + 6.994 + .72 =

9.
$$\begin{array}{r} 7.642 \\ 17.63 \\ 2.143 \\ +\ 14.64 \\ \hline \end{array}$$

10. 19.2 + 7.63 + 4.26 =

1.
2.
3.
4.
5.
6.
7.
8.
9.
10.
Score

Problem Solving

In January it rained 4.76 inches, in February, 6.43 inches, and in March, 7.43 inches. What was the total amount of rainfall for the three months?

Review Exercises

1. Write 7.19 in words.

2.   7.63
        4.2
      + 2.673

3. Write $7 \frac{5}{10,000}$ as a decimal.

4. Write 7.0632 as a mixed numeral.

5. Write .019 in words.

6. .7 + .4 + .8 =

Helpful Hints

Use what you have learned to solve the following problems.
* Zeros may be placed to the right of the decimal.
* Write the decimal point in the answer.

S1. 4.6	S2. 3.67 + 9.3 + .61 =
13.24	
+ 7.89	

1. 41.23 + 6.973 + 1.9 =

2. 5.912 + 5 + 7.63 =

3.   19.62
       7.426
    + 7.93

4. .36 + .79 + .6 =

5. 92.5 + 12 + 2.16 + .7 =

6.   73.197
       6.72
      17.109
    + 6.28

7. .9 + .6 + .8 =

8. 76.76 + 8.765 + .89 =

9.   9.726
     14.276
      7.93
   + 24.67

10. 17.6 + 3.68 + 4.27 =

1.
2.
3.
4.
5.
6.
7.
8.
9.
10.
Score

Problem Solving

Mr. Otis purchased milk for $3.15 and bread for $1.85. If he paid with a 20 dollar bill, what was the amount of change that he should receive?

Review Exercises

1. 7.24
    3.6
    + 8.717

2. 4.8 + 6 + 3.26 =

3. Write $7\frac{125}{10,000}$ as a decimal.

4. Write 6.012 as a mixed numeral.

5. Write $2\frac{7}{100}$ in words.

6. Write 7.005 in words.

Helpful Hints	To subtract decimals, line up the decimal points and subtract as you would whole numbers. Write the decimal points in the answer. Zeroes may be placed to the right of the decimal.	**Examples:** $3.2 - 1.66 =$ $\overset{2\;11\;1}{3.20}$ $-\ 1.66$ 1.54	$7 - 1.63 =$ $\overset{6\;9\;1}{7.00}$ $-\ 1.63$ 5.37

S1.   17.2
      - 3.36

S2.   15.1 - 7.62 =

1.   7.32
     - 1.426

2.   3.962
     - 1.669

3.   2.72 - 1.56 =

4.   27.93 - 16.8 =

5.   .72 - .667 =

6.   6.137
     - 2.1793

7.   3 - .627 =

8.   7.14 - 3.456 =

9.   75.6 - 66.972 =

10.   43.21 - 16.445 =

1.

2.

3.

4.

5.

6.

7.

8.

9.

10.

Score

Problem Solving

The normal temperature is 98.6°. If Gwen's temperature is 101.3°, how much above normal is her temperature?

Review Exercises

1. Which has the largest value, 2.6 or 2.599?

2. $6.23 + 2.7 + 3.24 =$

3. Write $\frac{72}{100}$ in words.

4. Write $\frac{72}{10,000}$ in words.

5. Write 75.0006 as a mixed numeral.

6. $15.8 + 75 + 6.23 =$

Helpful Hints	Use what you have learned to solve the following problems.
	  * Line up the decimal points.
	  * Zeroes may be added to the right of the decimal.
	  * Write the decimal point in the answer.

S1. $75.3 - 19.68 =$ S2. $75 - 1.96 =$

1.
$$\begin{array}{r} 6.35 \\ -\ 2.347 \\ \hline \end{array}$$

2.
$$\begin{array}{r} 7.547 \\ -\ .36 \\ \hline \end{array}$$

3. $7.15 - 3.672 =$ 4. $37.3 - .96 =$

5. $.89 - .697 =$

6.
$$\begin{array}{r} 7.136 \\ -\ 2.2476 \\ \hline \end{array}$$

7. $5 - .964 =$ 8. $95.1 - 7.124 =$

9. $6.2 - 3.914 =$ 10. $85.3 - 27.965 =$

1.	
2.	
3.	
4.	
5.	
6.	
7.	
8.	
9.	
10.	
Score	

Problem Solving	Mrs. Roberts had $210.25 in her savings account. On Monday she withdrew $56.75 and on Friday she made a deposit of $125.50. How much does she now have in her savings account?

Review Exercises

1. 7.63
    9.4
    + .76

2.   7.6
    - 2.542

3. $9.6 + 5 + 2.76 =$

4. $15 - 2.78 =$

5. Write $\dfrac{75}{1,000}$ as a decimal.

6. Write 2.058 in words.

Helpful Hints	Use what you have learned to solve the following problems.
	  * Line up the decimals.
	  * Put decimals in the answer.
	  * Zeroes may be added to the right of the decimal.

S1. 3.61
    14.4
    + .37

S2.   7.16
    - 3.473

1. 7.61
    8.92
    + 7.634

2.   7.6
    - 1.43

3. $4.63 + 5.7 + 6.24 =$

4. $17.2 - 8.96 =$

5. $15 - 12.92 =$

6. $6.93 + 5 + 7.63 =$

7. $.9 + .7 + .6 =$

8. $7.16 - 2.673 =$

9. $27.16 - 16.764 =$

10. $7.73 + 2.6 + .37 + 15 =$

1.
2.
3.
4.
5.
6.
7.
8.
9.
10.
Score

Problem Solving	In May Roberto weighed 145.2 pounds. In July he weighed 139.7 pounds. How much more did he weigh in May than in July?

Review Exercises

1. Write $\frac{7}{1,000,000}$ as a decimal.

2. Write $\frac{7}{1,000,000}$ in words.

3.
$$\begin{array}{r} 62.7 \\ 3.9 \\ 5. \\ +\ 7.62 \end{array}$$

4. Write six and twenty-two ten-thousandths as a decimal.

5. $7 - 2.36 =$

6.
$$\begin{array}{r} 5.1 \\ -\ 2.765 \end{array}$$

Helpful Hints

Use what you have learned to solve the following problems.

S1.
$$\begin{array}{r} 42.6 \\ .39 \\ +\ 6.427 \end{array}$$

S2.
$$\begin{array}{r} 9.14 \\ -\ 2.376 \end{array}$$

1. $7.62 + 5.97 + 3.6 =$

2.
$$\begin{array}{r} .9 \\ .7 \\ +\ .6 \end{array}$$

3. $27 - .27 =$

4. $4.9 - 1.666 =$

5. $36.19 + 24 + 32.916 =$

6. $4.963 - 2.9 =$

7. $.9 + 5 + .7 =$

8. $55.828 + 6.97 + 3.42 =$

9. $17.6 - 8.27 =$

10.
$$\begin{array}{r} 47.2 \\ 3.647 \\ 5.23 \\ +\ 16.924 \end{array}$$

1.
2.
3.
4.
5.
6.
7.
8.
9.
10.

Score

Problem Solving

The Jenkins family drove 1250.5 miles in three days. The first day they drove 550.25 miles and on the second day they drove 352.3 miles. How far did they drive on the third day?

Review Exercises

1. 34
    x 5

2. 47
    x 33

3. 307
    x 36

4. 7.63
    3.4
    + 9.66

5. 7.2
    - 2.637

6. 7.5 + 3 + 2.6 + 3.4 =

Helpful Hints

Multiply as you would with whole numbers. Find the number of decimal places and place the decimal point properly in the product.

Examples:

2.32 ←—2 places
x   6
13.92 ←—2 places

7.6 ←—1 place
x 23
228
1520
174.8 ←—1 place

S1. 2.46
     x   3

S2. 2.3
     x 16

1. .643
    x   3

2. 3.66
    x   4

3. .16
    x 43

4. 2.36
    x 24

5. 1.4
    x 16

6. 3.45
    x 16

7. 7.63
    x 43

8. 1.432
    x   7

9. .41
    x 73

10. .046
     x 27

1.
2.
3.
4.
5.
6.
7.
8.
9.
10.
Score

Problem Solving

A farmer can harvest 7.6 tons of crops per day.
How many tons of crops can be harvested in 5 days?

Review Exercises

1. 4.26
    x   3

2.   2.3
    x 14

3.   7.213
    x   5

4.   .032
    x   23

5.   7.1
    - 2.367

6.   7.6
     .72
     5.1
    + 6.327

Use what you have learned to solve the following problems.

* Remember to write the decimal in the answer in the proper location.

S1.   3.47
     x   5

S2.   4.4
     x 23

1.   .723
    x   4

2.   3.72
    x   6

3.   .27
    x 46

4.   3.62
    x 24

5.   4.2
    x 16

6.   3.45
    x 23

7.   7.124
    x   8

8.   2.97
    x   6

9.   7.5
    x 76

10.   .137
     x 24

1.
2.
3.
4.
5.
6.
7.
8.
9.
10.

Score

Problem Solving

Shirts are on sale for $20.15 each.
What would be the cost of five shirts?

Review Exercises

1. 7.25
    3.7
    + 4.637

2. 7.132
    - 1.476

3. 1.15
    x   5

4. .24
    x   16

5. 6 - .37 =

6. 7.21 + 9 + 6.426 =

Helpful Hints	Multiply as you would with whole numbers. Find the number of decimal places and place the decimal point properly in the product.	**Examples:**

2.63 ←—2 places
x   .3 ←—1 place
.789 ←—3 places

.724 ←—3 places
x .23 ←—2 places
2172
14480
.16652 ←—5 places

S1. 3.6
     x .7

S2. 3.24
     x 2.4

1. 3.6
    x 3.2

2. 2.09
    x .22

3. .642
    x .33

4. .23
    x 3.8

5. 2.03
    x .07

6. .422
    x 23.2

7. .003
    x 0.8

8. 5.6
    x 3.4

9. 63.5
    x 2.35

10. 12.3
     x .006

1.
2.
3.
4.
5.
6.
7.
8.
9.
10.

Score

Problem Solving

One pound of shrimp costs $3.20.
How much will 2.5 pounds cost?

Review Exercises

1. 6.42
    x   6

2. Write 3.0026 in words.

3. Write 3.007 as a mixed numeral.

4. Write two and eleven thousandths as a decimal.

5. Write $\dfrac{16}{100,000}$ in words.

6.   4.6
    x .23

Helpful Hints

Use what you have learned to solve the following problems.

* Remember to place the decimal properly in the answer.

1.	
2.	
3.	
4.	
5.	
6.	
7.	
8.	
9.	
10.	
Score	

S1.   .324
    x   .7

S2.   3.26
    x 4.2

1.   5.3
    x 4.6

2.   2.09
    x .33

3.   .426
    x .44

4.   4.07
    x .06

5.   .27
    x 4.6

6.   .433
    x 2.73

7.   .007
    x 0.6

8.   6.4
    x .43

9.   6.51
    x 2.34

10.   12.7
    x .008

Problem Solving

Five friends are going to attend a concert together.
If tickets are $17.50 each, what is the total cost of tickets?

Review Exercises

1. 2.14
    x   3

2. 2.13
    x   1.7

3. .002
    x .003

4. 3.63
    7.75
    + 4.62

5. 5.1
    - 3.243

6. $3.54
    x   5

Helpful Hints	To multiply by 10, move the decimal point one place to the right; by 100, two places to the right; and by 1,000, three places to the right

Examples:
10 x 3.36 = 33.6
100 x 3.36 = 336
1000 x 3.36 = 3360*
*Sometimes placeholders are necessary.

S1. 10 x 3.2 =

S2. 1,000 x 7.39 =

1. 100 x .936 =

2. 1,000 x 72.6 =

3. 100 x 1.6 =

4. 7.362 x 100 =

5. 7.28 x 1000 =

6. 100 x .7 =

7. 100 x .376 =

8. 1000
    x   .39

9. 100 x .733 =

10. 10 x 7.63 =

1.
2.
3.
4.
5.
6.
7.
8.
9.
10.
Score

Problem Solving

If tickets to a concert are $18.50, how much would 1,000 tickets cost?

Review Exercises

1. Write 2.763 in words.

2. Write seven hundred-thousandths as a decimal

3. $5 - .55 =$

4. Write $\dfrac{16}{100,000}$ as a decimal.

5. $\begin{array}{r} .723 \\ \times \quad .5 \\ \hline \end{array}$

6. $7.6 + 44.27 + 1.93 =$

Helpful Hints	Use what you have learned to solve the following problems. * Sometimes place holders are necessary

S1. $100 \times 3.7 =$ S2. $1,000 \times 5.36 =$

1. $100 \times 3.27 =$ 2. $1,000 \times 56.7 =$

3. $100 \times 1.9 =$ 4. $7.364 \times 100 =$

5. $.976 \times 1000 =$ 6. $1,000 \times .75 =$

7. $10 \times 7.3 =$ 8. $\begin{array}{r} 1000 \\ \times \quad .387 \\ \hline \end{array}$

9. $100 \times 83.3 =$ 10. $8.42 \times 10 =$

1.
2.
3.
4.
5.
6.
7.
8.
9.
10.
Score

Problem Solving	If individual floor tiles weigh 2.5 pounds each, what is the weight of 1,000 floor tiles?

Review Exercises

1. 23.2
    x   6

2.   .36
    x   25

3.   32.5
    x 100

4.   7.06
    x .6

5.   .003
    x .003

6.   4.23
    x   5.1

Helpful Hints	Use what you have learned to solve the following problems. * Be careful when placing the decimal point in the product. * Sometimes place holders are necessary.

S1.   .342
    x   7

S2.   42.3
    x   .36

1.   .23
    x   14

2.   .29
    x 1.6

3.   1.34
    x .362

4. $10 \times 2.6 =$

5.   2.63
    x 1.2

6. $100 \times 26.3 =$

7.   .003
    x   3.6

8.   .65
    x 5.5

9.   1.67
    x   33

10.   .67
    x .063

1.

2.

3.

4.

5.

6.

7.

8.

9.

10.

Score

Problem Solving	If Omar earns $16 per hour, how much will he earn in 4.5 hours?

Review Exercises

1. 2.76 + 5 + 3.99 =   2. 7.1 - 2.334 =   3.   62.83
                                                                 7.9
                                                            +   4.652

4. 15 - 2.76 =   5. 15 - .12 =   6.   2.365
                                                            x   1000

Helpful Hints	Use what you have learned to solve the following problems. * Be careful when placing the decimal point in the product. * Practice reading the answers to yourself.

S1.   4.26   S2.   .062   1.   42
    x   6   x   5.8   x   2.3

2.   .47   3.   2.09   4. 100 x 3.6 =
    x   6.3   x   .003

5.   37.2   6.   .005   7.   26.4
    x   2.4   x   .001   x   5.23

8.   42.9 x 1,000 = 9.   2.17   10.   .046
                         x   16   x   0.63

1.
2.
3.
4.
5.
6.
7.
8.
9.
10.
Score

Problem Solving	A man bought three bags of chips at $.59 each and a pizza for $11.75, how much did he spend in all?

159

Review Exercises

1. $6\overline{)134}$

2. $5\overline{)1515}$

3. $22\overline{)2442}$

4. $22\overline{)7963}$

5. $18\overline{)2376}$

6. $25\overline{)5075}$

| **Helpful Hints** | Divide as you would with whole numbers. Place the decimal point directly up. | **Examples:** | $$\begin{array}{r} 2.8 \\ 3\overline{)8.4} \\ -6\downarrow \\ \hline 24 \\ -24 \\ \hline 0 \end{array}$$ | $$\begin{array}{r} .084 \\ 3\overline{).252} \\ -24\downarrow \\ \hline 12 \\ -12 \\ \hline 0 \end{array}$$ |

S1. $3\overline{)1.32}$

S2. $8\overline{)14.4}$

1. $3\overline{)59.1}$

2. $7\overline{)22.47}$

3. $34\overline{)19.38}$

4. $70.3 \div 19 =$

5. $4\overline{)24.32}$

6. $6\overline{)245.4}$

7. $26\overline{)8.424}$

8. $16\overline{)2.56}$

9. $12.72 \div 6 =$

10. $21\overline{)42.84}$

1.

2.

3.

4.

5.

6.

7.

8.

9.

10.

Score

| **Problem Solving** | If five loaves of bread cost $9.75, how much does one loaf cost? |

Review Exercises

1. $3\overline{)13.2}$

2. $5\overline{)1.725}$

3. $22\overline{)35.86}$

4. $\begin{array}{r} 7.12 \\ -\ 1.637 \\ \hline \end{array}$

5. $\begin{array}{r} .36 \\ \times\ 2.4 \\ \hline \end{array}$

6. $\begin{array}{r} 36.8 \\ 4.92 \\ +\ 36.57 \\ \hline \end{array}$

Helpful Hints

Use what you have learned to solve the following problems.
* Place the decimal point directly up.

S1. $4\overline{).928}$ S2. $7\overline{)18.76}$ 1. $3\overline{)1.53}$

2. $8\overline{)32.48}$ 3. $32\overline{)17.92}$ 4. $41\overline{)229.6}$

5. $3\overline{)6.912}$ 6. $65\overline{)14.95}$ 7. $24\overline{)15.12}$

8. $32\overline{)5.12}$ 9. $6\overline{).738}$ 10. $21\overline{)85.68}$

1.	
2.	
3.	
4.	
5.	
6.	
7.	
8.	
9.	
10.	
Score	

Problem Solving

Al's times for the one hundred yard dash were 11.8 seconds, 12.5 seconds, and 12.3 seconds. What was his average time?

Review Exercises

1. $3\overline{)22.14}$

2. $15\overline{)4.515}$

3. $100 \times 2.7 =$

4. $75 - 36.2 =$

5. Find the sum of 256.7 and 22.96.

6. Write 2.1762 as a mixed numeral.

Helpful Hints	Sometimes placeholders are necessary when dividing decimals.	**Examples:** $\begin{array}{r} .05 \\ 3\overline{)15} \\ -15 \\ \hline 0 \end{array}$ $\begin{array}{r} .003 \\ 15\overline{)045} \\ -45 \\ \hline 0 \end{array}$

S1. $5\overline{).0135}$ S2. $13\overline{).247}$ 1. $7\overline{).0049}$

2. $3\overline{).036}$ 3. $4\overline{).224}$ 4. $13\overline{).468}$

5. $22\overline{).946}$ 6. $9\overline{).567}$ 7. $52\overline{)1.196}$

8. $18\overline{).396}$ 9. $9\overline{).027}$ 10. $12\overline{).816}$

1.

2.

3.

4.

5.

6.

7.

8.

9.

10.

Score

Problem Solving	A man bought two hammers for $5.25 each and three saws for $7.50 each. What was the total cost?

Review Exercises

1. Write 2.003 in words.

2.
$$\begin{array}{r} 32.6 \\ 5. \\ + \ 72.613 \\ \hline \end{array}$$

3.
$$\begin{array}{r} 76.2 \\ - \ 9.213 \\ \hline \end{array}$$

4.
$$\begin{array}{r} 62.3 \\ \times \quad 7 \\ \hline \end{array}$$

5. $4\overline{)53.2}$

6. $22\overline{)3.784}$

Helpful Hints	Use what you have learned to solve the following problems.
	* Be careful to include placeholders when necessary.

S1. $6\overline{).330}$ S2. $13\overline{).494}$ 1. $9\overline{).0036}$

2. $7\overline{).014}$ 3. $4\overline{).288}$ 4. $12\overline{).552}$

5. $42\overline{)1.428}$ 6. $6\overline{).324}$ 7. $62\overline{)3.534}$

8. $18\overline{).792}$ 9. $8\overline{).0064}$ 10. $12\overline{).408}$

1.
2.
3.
4.
5.
6.
7.
8.
9.
10.
Score

Problem Solving

Shirts are on sale for $9.50 each. If the regular price is $11.25, how much would be saved by buying three shirts on sale?

Review Exercises

1.   22.6
     x   3

2.   .342
     x .06

3.   .36 x 1,000 =

4.   5⟌.015

5.   12⟌.132

6.   Write six and fifteen hundred-thousandths as a decimal.

Helpful Hints	Sometimes zeroes need to be added to the dividend to complete the problem. * Sometimes it is necessary to add more than one zero.	**Examples:** $5⟌\overline{1.3}$ $15⟌\overline{2.7}$	$\begin{array}{r}.26\\5⟌\overline{1.30}\\-1\,0\!\downarrow\\\hline 30\\-30\\\hline 0\end{array}$	$\begin{array}{r}.18\\15⟌\overline{2.70}\\-1\,5\!\downarrow\\\hline 120\\-120\\\hline 0\end{array}$

S1.   5⟌1.7

S2.   25⟌1.5

1.   2⟌.13

2.   5⟌3.1

3.   22⟌45.1

4.   24⟌3.6

5.   5⟌0.2

6.   95⟌3.8

7.   20⟌2.4

8.   4⟌6.3

9.   5⟌.03

10.   5⟌2.09

1.

2.

3.

4.

5.

6.

7.

8.

9.

10.

Score

Problem Solving

A woman bought three gallons of gas at $3.29 per gallon. If she paid with a $20.00 bill, what would be the change?

Review Exercises

1. $5\overline{)\$7.95}$

2. $\begin{array}{r} \$7.90 \\ \times\ \ \ \ 8 \\ \hline \end{array}$

3. $2\overline{).13}$

4. $42.6 + 3.92 + .96 =$

5. $55.23 - 36.712 =$

6. $95\overline{)7.6}$

Helpful Hints	Use what you have learned to solve the following problems.
	  * Use placeholders when necessary.
	  * Add zeroes to the dividend when necessary.

S1. $5\overline{).37}$ S2. $22\overline{)49.5}$ 1. $2\overline{).17}$

2. $5\overline{).39}$ 3. $20\overline{)1.2}$ 4. $5\overline{).8}$

5. $5\overline{)3.19}$ 6. $18\overline{)2.7}$ 7. $4\overline{)5.8}$

8. $2\overline{).37}$ 9. $4\overline{)5.4}$ 10. $15\overline{).6}$

1.	
2.	
3.	
4.	
5.	
6.	
7.	
8.	
9.	
10.	
Score	

Problem Solving	If 3 pounds of butter is $5.43, what is the price per pound?

Review Exercises

1. $2\overline{).19}$

2. $5\overline{)1.7}$

3. $6\overline{).0012}$

4. Change $5\frac{13}{10,000}$ to a decimal.

5. Write .00135 as a fraction.

6.   .7
        .9
    +   .6
    ‾‾‾‾‾‾

Helpful Hints

When dividing a decimal by another decimal, move the decimal point in the divisor the number of places necessary to make it a whole number. Move the decimal point in the dividend the same number of places.

Examples:

$$.3\overline{)2.4}\quad\begin{array}{r}.8\\-2\,4\\\hline 0\end{array}$$

$$.03\overline{)28.50}\quad\begin{array}{r}9\,50.*\\-27\downarrow\\\hline 1\,5\\-1\,5\\\hline 0\end{array}$$

*Sometimes placeholders are necessary

S1. $.7\overline{)2.73}$

S2. $.15\overline{).036}$

1. $.3\overline{)2.4}$

2. $.03\overline{)5.1}$

3. $.9\overline{).378}$

4. $.04\overline{)3.2}$

5. $.06\overline{).324}$

6. $2.1\overline{)6.72}$

7. $.26\overline{).962}$

8. $.18\overline{).576}$

9. $.04\overline{)2.3}$

10. $.12\overline{)1.104}$

1.	
2.	
3.	
4.	
5.	
6.	
7.	
8.	
9.	
10.	
Score	

Problem Solving

Five cans of beans cost $2.75. One can of tuna costs $.79. How much would it cost for one can of beans and two cans of tuna?

Review Exercises

1. 0.23 + 1.47 + 0.37 =

2. 8.7 - 2.79 =

3. 1.2 x 3.42 =

4. 1.3 ÷ 2 =

5. .17 ÷ 5 =

6. $12\overline{)5.16}$

Helpful Hints	Use what you have learned to solve the following problems.
	* Use placeholders when necessary.
	* Add zeroes to the dividend when necessary.

S1. $.8\overline{)2.5}$

S2. $.15\overline{)76.2}$

1. $.2\overline{).25}$

2. $.8\overline{).024}$

3. $.15\overline{).762}$

4. $.04\overline{)9.2}$

5. $.02\overline{).4}$

6. $.05\overline{).325}$

7. $2.1\overline{).693}$

8. $.15\overline{)7.5}$

9. $.53\overline{)4.876}$

10. $.03\overline{).072}$

1.

2.

3.

4.

5.

6.

7.

8.

9.

10.

Score

Problem Solving	A stack of tiles is 27.5 inches tall. If each tile is .5 inches tall, how many tiles are in the stack?

Review Exercises

1. $3\overline{)4.8}$ 2. $.04\overline{).6}$ 3. $.5\overline{).17}$

4. $.12\overline{)36}$ 5. $.29\overline{)2.929}$ 6. $1.3\overline{)6.76}$

Helpful Hints

To change fractions to decimals, divide the numerator by the denominator. Add as many zeroes as necessary.

Examples:

$\dfrac{3}{4}$ $4\overline{)3.00}$ $.75$
$-2\ 8$
-20
-20
0

$\dfrac{3}{8}$ $8\overline{)3.000}$ $.375$
-24
60
-56
40
-40
0

Change each fraction to a decimal.

S1. $\dfrac{3}{4}$ S2. $\dfrac{5}{8}$ 1. $\dfrac{3}{5}$

2. $\dfrac{1}{4}$ 3. $\dfrac{2}{5}$ 4. $\dfrac{7}{8}$

5. $\dfrac{11}{20}$ 6. $\dfrac{13}{25}$ 7. $\dfrac{5}{8}$

8. $\dfrac{4}{20}$ 9. $\dfrac{1}{5}$ 10. $\dfrac{7}{10}$

1.

2.

3.

4.

5.

6.

7.

8.

9.

10.

Score

Problem Solving

A worker earned $700 and put .6 of it into his savings account. How much did he put into his savings account?
(Hint: What operation does "of" usually mean?)

Review Exercises

1. Which is the larger decimal, 2.3 or 2.199?

2.
$$7.73$$
$$14.2$$
$$+ \ 7.16$$

3.
$$.207$$
$$\text{x} \ \ 1.4$$

4. $5 - .123 =$

5. $.5\overline{)2}$

6. $.03\overline{)1.5}$

Helpful Hints	Use what you have learned to solve the following problems. * Add as many zeroes as necessary.

Change each fraction to a decimal.

S1. $\dfrac{3}{8}$	S2. $\dfrac{3}{4}$	1. $\dfrac{4}{5}$
2. $\dfrac{3}{16}$	3. $\dfrac{11}{25}$	4. $\dfrac{3}{20}$
5. $\dfrac{6}{8}$	6. $\dfrac{1}{5}$	7. $\dfrac{13}{50}$
8. $\dfrac{7}{10}$	9. $\dfrac{7}{25}$	10. $\dfrac{7}{16}$

1.

2.

3.

4.

5.

6.

7.

8.

9.

10.

Score

Problem Solving	Cans of peas are on sale for two for $.69. How much would eight cans cost?

Review Exercises

1. $6\overline{)134}$

2. $5\overline{)1515}$

3. $22\overline{)2442}$

4. $22\overline{)7963}$

5. $18\overline{)2376}$

6. $25\overline{)5075}$

Helpful Hints	Divide as you would with whole numbers. Place the decimal point directly up.	Examples: $\begin{array}{r} 2.8 \\ 3\overline{)8.4} \\ -6\downarrow \\ \hline 24 \\ -24 \\ \hline 0 \end{array}$ $\begin{array}{r} .084 \\ 3\overline{).252} \\ -24\downarrow \\ \hline 12 \\ -12 \\ \hline 0 \end{array}$

S1. $3\overline{)1.32}$ S2. $8\overline{)14.4}$ 1. $3\overline{)59.1}$

2. $7\overline{)22.47}$ 3. $34\overline{)19.38}$ 4. $70.3 \div 19 =$

5. $4\overline{)24.32}$ 6. $6\overline{)245.4}$ 7. $26\overline{)8.424}$

8. $16\overline{)2.56}$ 9. $12.72 \div 6 =$ 10. $21\overline{)42.84}$

1.

2.

3.

4.

5.

6.

7.

8.

9.

10.

Score

Problem Solving	If five loaves of bread cost $9.75, how much does one loaf cost?

Review Exercises

1. $3\overline{)59.1}$ 2. $16\overline{)2.56}$ 3. $5\overline{).015}$

4. $4\overline{)6.3}$ 5. $.06\overline{)3.24}$ 6. $.21\overline{).672}$

Helpful Hints	Use what you have learned to solve the following problems. * If necessary, refer to the "Helpful Hints" section from previous pages.

S1. $2\overline{)1.25}$ S2. $.07\overline{)4.9}$ 1. $5\overline{).039}$

1.	
2.	
3.	
4.	
5.	
6.	
7.	
8.	
9.	
10.	
Score	

2. $3\overline{)2.34}$ 3. $.02\overline{).146}$ 4. $.24\overline{)1.2}$

5. $.3\overline{)6.375}$ 6. $.005\overline{)1.43}$ 7. $4.5\overline{)1.035}$

8. $.84\overline{).546}$ 9. Change $\frac{3}{4}$ to a decimal.   10. Change $\frac{1}{4}$ to a decimal.

Problem Solving	A man bought a car. He made a down payment of $1,200.00 and paid $300 per month for 36 months. How much did he pay altogether?

Final Review of All Decimal Operations

1. 4.56
    7.8
    + 3.976

2. .3 + 4.67 + 8.9 =

3. 16.8 + 5 + 12.7 =

4. 47.6
    - 19.7

5. 9.2
    - 3.652

6. 72 - 1.49 =

7. 4.76
    x   4

8. 5.6
    x   23

9. .49
    x 2.6

10. .503
     x 3.46

11. 1,000 x 3.19 =

12. 100 x 3.2 =

13. $5\overline{).79}$

14. $2\overline{)3.96}$

15. $.004\overline{)1.2}$

16. $.7\overline{).224}$

17. $.15\overline{).0045}$

18. $6.8\overline{)16.32}$

19. Change $\frac{5}{8}$ to a decimal.

20. Change $\frac{9}{16}$ to a decimal.

1.
2.
3.
4.
5.
6.
7.
8.
9.
10.
11.
12.
13.
14.
15.
16.
17.
18.
19.
20.
Score

Final Test of all Decimal Operations

1.
```
     5.62
    15.7
     8.236
  + 12.16
```

2. $7.6 + 5 + .9 + 2.72 =$

3. $.7 + .6 + .9 + .7 =$

4. $72 - .72 =$

5.
```
    72.6
  - 19.723
```

6. $.3 - .216 =$

7.
```
    .937
  x    5
```

8.
```
    15
  x 5.6
```

9.
```
    .72
  x 4.9
```

10.
```
    .207
  x  .69
```

11. $4.1 \times 100 =$

12. $3.762 \times 1{,}000 =$

13. $5\overline{)3.95}$

14. $5\overline{).032}$

15. $.9\overline{).225}$

16. $.005\overline{)30}$

17. $.25\overline{)117.5}$

18. $.12\overline{).2544}$

19. Change $\frac{1}{8}$ to a decimal.

20. Change $\frac{7}{8}$ to a decimal.

1.	
2.	
3.	
4.	
5.	
6.	
7.	
8.	
9.	
10.	
11.	
12.	
13.	
14.	
15.	
16.	
17.	
18.	
19.	
20.	
Score	

Review Exercises

1. Change $\dfrac{17}{100}$ to a decimal.   2. Change $\dfrac{9}{10}$ to a decimal.   3. Write .07 as a fraction.

4. Write .7 as a fraction.   5. Write $6\dfrac{7}{100}$ in words.   6. Write 10.9 in words.

Helpful Hints	Percent means "per hundred" or "hundredths." If a fraction is expressed as hundredths, it can easily be written as a percent.	**Examples:** $\dfrac{7}{100} = 7\%$ $\dfrac{3}{10} = 30\%$ $\dfrac{19}{100} = 19\%$

Change each of the following to a percent.

S1. $\dfrac{17}{100} =$ S2. $\dfrac{9}{10} =$ 1. $\dfrac{6}{100} =$

2. $\dfrac{99}{100} =$ 3. $\dfrac{3}{10} =$ 4. $\dfrac{64}{100} =$

5. $\dfrac{67}{100} =$ 6. $\dfrac{1}{100} =$ 7. $\dfrac{7}{10} =$

8. $\dfrac{14}{100} =$ 9. $\dfrac{80}{100} =$ 10. $\dfrac{62}{100} =$

1.
2.
3.
4.
5.
6.
7.
8.
9.
10.

Score

Problem Solving	Ron is taking a trip of 252 miles. If his car gets 21 miles per gallon of gas, how many gallons of gas will the car consume? If gas costs \$3.25 per gallon, how much will Ron spend on gas for the entire trip?

Review Exercises

1. $7\overline{)\,.119}$

2. $.003\overline{)\,1.5}$

3. $100 \times 3.4 =$

4. $\begin{array}{r} 6.12 \\ \times\ .7 \\ \hline \end{array}$

5. $7.63 + 52 + 9.64 =$

6. $\begin{array}{r} 7.1 \\ -\ 2.964 \\ \hline \end{array}$

Helpful Hints

Use what you have learned to solve the following problems.

Change each of the following to a percent.

S1. $\dfrac{7}{10} =$	S2. $\dfrac{3}{100} =$	1. $\dfrac{19}{100} =$

2. $\dfrac{87}{100} =$ 3. $\dfrac{6}{10} =$ 4. $\dfrac{63}{100} =$

5. $\dfrac{19}{100} =$ 6. $\dfrac{2}{100} =$ 7. $\dfrac{48}{100} =$

8. $\dfrac{14}{100} =$ 9. $\dfrac{5}{10} =$ 10. $\dfrac{98}{100} =$

1. _____
2. _____
3. _____
4. _____
5. _____
6. _____
7. _____
8. _____
9. _____
10. _____

Score _____

Problem Solving

Ellen worked 25 hours and earned 8 dollars per hour. If she wants to buy a bike for $372, how much more money does she need?

Review Exercises

1. $\dfrac{7}{10}$ = _____%

2. $\dfrac{72}{100}$ = _____%

3. $.03\overline{)12}$

4. Change $\dfrac{5}{8}$ to a decimal.

5. Change $\dfrac{3}{5}$ to a decimal.

6. $\begin{array}{r} .003 \\ \times\ .002 \\ \hline \end{array}$

Helpful Hints	"Hundredths" = Percent Decimals can easily be changed to percents.	**Examples:** .27 = 27% .9 = .90 = 90% .124 = 12.4% * Move the decimal point twice to the right and add a percent symbol.

Change each of the following to a percent.

S1. .37

S2. .7

1. .93

2. .02

3. .2

4. .09

5. .6

6. .665

7. .89

8. .6

9. .334

10. .8

1.

2.

3.

4.

5.

6.

7.

8.

9.

10.

Score

Problem Solving	There are 16 fluid ounces in a pint. How many fluid ounces are there in .6 pints?

Review Exercises

1. Write 3.0196 as a mixed numeral.

2. Write .0021 in words.

3. Write $7\frac{7}{10,000}$ as a decimal.

4. $.15\overline{)\ .4545}$

5. $.5\overline{)\ 2\ }$

6. $1.6\overline{)\ .352}$

| **Helpful Hints** | Use what you have learned to solve the following problems.
* Move the decimal twice to the right and add a percent symbol. |

Change each of the following to a percent.

S1. .09	S2. .348	1. .90	1. _____
			2. _____
			3. _____
2. .09	3. .7	4. .097	4. _____
			5. _____
			6. _____
5. .6	6. .007	7. .87	7. _____
			8. _____
			9. _____
8. .3	9. .445	10. .4	10. _____
			Score

Problem Solving   A school has 480 students. If .25 of them ride the bus to school, how many students take the bus? (Hint: What does "of" mean?)

Review Exercises

1. 7.9 - .79 =

2. 5 - 2.78 =

3. 3.46 + 15 + .78 =

4. .9 + .76 + .73 + .8 =

5.   5.13
    - 2.667

6.   17.54
        6.723
    + 36.124

Helpful Hints	Percents can be expressed as decimals and fractions. The fraction form may sometimes be reduced to its lowest terms.	**Examples:** $25\% = .25 = \dfrac{25}{100} = \dfrac{1}{4}$ $8\% = .08 = \dfrac{8}{100} = \dfrac{2}{25}$

Change each percent to a decimal and to a fraction reduced to its lowest terms.

S1.   20% = .   = ___

S2.   9% = .   = ___

1.   16% = .   = ___

2.   6% = .   = ___

3.   75% = .   = ___

4.   40% = .   = ___

5.   1% = .   = ___

6.   45% = .   = ___

7.   12% = .   = ___

8.   5% = .   = ___

9.   50% = .   = ___

10.   13% = .   = ___

1.
2.
3.
4.
5.
6.
7.
8.
9.
10.
Score

Problem Solving	If 25% of the students at Eaton School take the bus, what fraction of the students take the bus? (Reduce your answer to its lowest terms.)

Review Exercises

1.   12.7
    x   5

2.   24.5
    x .75

3.   .008
    x .03

4. $2\overline{).15}$

5. $1.2\overline{).2424}$

6. $.18\overline{).468}$

Helpful Hints	Use what you have learned to solve the following problems. * Be sure fractions are reduced to lowest terms.

Change each percent to a decimal and to a fraction reduced to its lowest terms.

S1.   50% = .   = ___

S2.   5% = .   = ___

1.   8% = .   = ___

2.   80% = .   = ___

3.   24% = .   = ___

4.   11% = .   = ___

5.   2% = .   = ___

6.   70% = .   = ___

7.   9% = .   = ___

8.   90% = .   = ___

9.   17% = .   = ___

10.   14% = .   = ___

1.

2.

3.

4.

5.

6.

7.

8.

9.

10.

Score

Problem Solving	A yard is in the shape of a rectangle that is 10 feet wide and 12 feet long. To build a fence it costs $20 per foot. How much would it cost to build a fence around the yard? (Hint: Make a sketch.)

Review Exercises

1. Change 80% to a decimal.

2. Change 7% to a decimal.

3. Change 25% to fraction reduced to lowest terms.

4.
$$156 \times .7$$

5.
$$400 \times .32$$

6.
$$300 \times .06$$

Helpful Hints

To find the percent of a number you may use either fractions or decimals. Use what is the most convenient.

Examples: Find 25% of 60

.25 x 60

$$\begin{array}{r} 60 \\ \times .25 \\ \hline 300 \\ 120 \\ \hline 15.00 \end{array}$$

OR

$$\frac{25}{100} = \frac{1}{4}$$

$$\frac{1}{4_1} \times \frac{\overset{15}{\cancel{60}}}{1} = \frac{15}{1} = 15$$

S1. Find 70% of 25.

S2. Find 50% of 300.

1. Find 6% of 72.

2. Find 60% of 85.

3. Find 25% of 60.

4. Find 45% of 250.

5. Find 10% of 320.

6. Find 40% of 200.

7. Find 4% of 250.

8. Find 90% of 240.

9. Find 75% of 150.

10. Find 2% of 660.

1.	
2.	
3.	
4.	
5.	
6.	
7.	
8.	
9.	
10.	
Score	

Problem Solving

A train traveled 400 miles in 2.5 hours. What was its average speed per hour?

Review Exercises

1. $.12\overline{).048}$

2. $5\overline{)2}$

3. $.2\overline{).13}$

4. Find .9 of 45

5. Write $\frac{3}{8}$ as a decimal.

6. Write 3.0016 in words.

Helpful Hints

Use what you have learned to solve the following problems.
* Use fractions or decimals, depending on which is the most convenient.

S1. Find 6% of 400.

S2. Find 60% of 400.

1. Find 4% of 80.

2. Find 70% of 550.

3. Find 7% of 550.

4. Find 25% of 200.

5. Find 20% of 250.

6. Find 60% of 310.

7. Find 5% of 220.

8. Find 80% of 300.

9. Find 75% of 160.

10. Find 25% of 24.

1.
2.
3.
4.
5.
6.
7.
8.
9.
10.
Score

Problem Solving

Rosa's test scores were 80, 90, 86, and 80.
What was her average test score?

Review Exercises

1. Find 12% of 60.

2. Find 90% of 320.

3. Find 25% of 40.

4. Change $\frac{3}{4}$ to a decimal.

5. Change 50% to a fraction reduced to the lowest terms.

6. Find .3 of 75.

Helpful Hints

When finding the percent of a number in a word problem, you can change the percent to a fraction or a decimal. Always express your answer in a short phrase or sentence.

Example:

A team played 60 games and won 75% of them. How many games did they win?

Find 75% of 60.

.75 x 60

$$\begin{array}{r} 60 \\ \times\ .75 \\ \hline 300 \\ 420 \\ \hline 45.00 \end{array}$$

OR

$$\frac{75}{100} = \frac{3}{4}$$

$$\frac{3}{4} \times \frac{60^{15}}{1} = \frac{45}{1} = 45$$

Answer: The team won 45 games.

S1. George took a test with 40 problems. If he got 15% of the problems correct, how many problems did he get correct?

S2. If 6% of the 500 students enrolled in a school are absent, then how many students are absent?

1. A worker earned 120 dollars and put 70% of it into the bank. How many dollars did he put into the bank?

2. A car costs $9,000. If Mr. Smith has saved 30% of this amount, how much did he save?

3. Steve took a test with 60 problems. If he got 70% of the problems correct how many of the problems did he get incorrect?

4. A family's monthly income is $3,000. If 20% of this amount is spent on food, how much money is spent on food?

5. There are 30 students in a class. If 60% of the class is boys, how many girls are in the class?

6. A house that costs $200,000 requires a 20% down payment. How many dollars are required for the down payment?

7. If a car costs $8,000 and loses 30% of its value in one year, how much will the car be worth in one year?

8. A coat is priced $50. If the sales tax is 7% of the price, how much is the sales tax? What is the total cost including sales tax?

9. 25% of the 600 students at Madison School take instrumental music. How many students are taking instrumental music?

10. A family spends 20% of its income for food and 30% for housing. If its monthly income is $3,000, how much is spent each month on food and housing?

1.
2.
3.
4.
5.
6.
7.
8.
9.
10.
Score

Problem Solving

A man had $362.00 in the bank. On Monday he withdrew $92.00, on Tuesday he deposited $76.00, and on Wednesday he withdrew $49.00. How much does he now have in the bank?

Review Exercises

1. Find 20% of 240.

2. Find 2% of 240.

3. Find $\frac{3}{4}$ of 240.

4. Change $\frac{1}{5}$ to a decimal.

5.
$$\begin{array}{r} .9 \\ .6 \\ + .8 \\ \hline \end{array}$$

6. 7 - 5.55 =

Helpful Hints

Use what you have learned to solve the following problems.
* Use fractions or decimals depending on which is most convenient.
* Express your answer in a short phrase or sentence.

S1. 40 people in a class take a test. If 80% passed the test, how many passed?

S2. In a class of 30 people 40% are boys. How many are girls?

1. A bakery made 600 cookies and sold 90% of them. How many cookies were sold?

2. A bag has a mixture of 120 white and red marbles. If 40% of the marbles are red, how many white marbles are there?

3. In a school of 800 students, 60% ride the bus. How many ride the bus?

4. A book costs $12.00. If it was on sale for 30% off, how much would be saved buying it on sale?

5. Marco is buying a car priced at $12,000. If he needs a down payment of 30%, how much is the down payment?

6. Steve earns $600 and saves 20% of it. How much does he spend?

7. Allie's bill at a restaurant was $45.00. If she wanted to leave a 20% tip, how much should she leave?

8. Bill earns $800. If 40% goes to rent and 20% goes to his car payment, what is the total cost for his rent and car payment?

9. 150 math students took a test and 80% passed. How many did not pass?

10. A school has 600 student and 60% are boys. How many boys are there in the school? How many girls?

1.
2.
3.
4.
5.
6.
7.
8.
9.
10.
Score

Problem Solving

Two pounds of beef costs $2.50. How much does six pounds cost?

Review Exercises

1. $.05\overline{).245}$

2. Find 20% of 36.

3. Find 2% of 360.

4. $27.2 - 18.76 =$

5. Change $7\frac{9}{1,000}$ to a decimal.

6. Change $\frac{9}{100}$ to a decimal.

Helpful Hints

To change a fraction to a percent, first change the fraction to a decimal, then change the decimal to a percent. Move the decimal twice to the right and add a percent symbol.

Examples:

$$\frac{3}{4} \qquad \begin{array}{r} .75 = 75\% \\ 4\overline{)3.00} \\ -2.8\downarrow \\ \hline -20 \\ -20 \\ \hline 0 \end{array}$$

$$\frac{16}{20} = \frac{4}{5} \qquad \begin{array}{r} .80 = 80\% \\ 5\overline{)4.00} \\ -4.0 \\ \hline 0 \end{array}$$

* Sometimes the fraction can be reduced further.

Change each of the following to a percent.

S1. $\frac{1}{5} =$

S2. $\frac{12}{15} =$

1. $\frac{3}{5} =$

2. $\frac{1}{2} =$

3. $\frac{1}{10} =$

4. $\frac{9}{12} =$

5. $\frac{15}{20} =$

6. $\frac{15}{25} =$

7. $\frac{1}{4} =$

8. $\frac{24}{30} =$

9. $\frac{18}{24} =$

10. $\frac{4}{20} =$

1.	
2.	
3.	
4.	
5.	
6.	
7.	
8.	
9.	
10.	

Problem Solving

Three hundred sixty people work for a company. Forty percent of them carpool to work. Find how many people carpool to work.

Score

Review Exercises

1.
   33.3
   4.44
   + 55.55

2. 15 - .15 =

3.
   2.17
   x   .7

4. 5)‾4

5. .07)‾.777

6. .14)‾.0294

Helpful Hints	Use what you have learned to solve the following problems. * Some fractions can be reduced further.

Change each of the following to a percent.

S1. $\dfrac{20}{25}$ =

S2. $\dfrac{45}{60}$ =

1. $\dfrac{4}{8}$ =

2. $\dfrac{7}{10}$ =

3. $\dfrac{6}{24}$ =

4. $\dfrac{5}{25}$ =

5. $\dfrac{5}{20}$ =

6. $\dfrac{4}{16}$ =

7. $\dfrac{5}{8}$ =

8. $\dfrac{30}{60}$ =

9. $\dfrac{9}{36}$ =

10. $\dfrac{3}{8}$ =

1.
2.
3.
4.
5.
6.
7.
8.
9.
10.

Score

Problem Solving	A worker completed $\dfrac{15}{20}$ of his project. What percent of his project has been completed?

Review Exercises

1. Find 15% of 310.

2. Find 20% of 120.

3. $8\overline{)6}$

4. Change $\frac{1}{2}$ to a percent.

5. Find .9 of 150.

6. $.05\overline{)30}$

Helpful Hints

When finding the percent, first write a fraction, change the fraction to a decimal, then change the decimal to a percent.   * "Is" means =.

Examples:

4 is what percent of 16?

$\frac{4}{16} = \frac{1}{4}$

$\begin{array}{r} .25 = 25\% \\ 4\overline{)1.00} \\ -8 \\ \hline -20 \\ -20 \\ \hline 0 \end{array}$

5 is what percent of 25?

$\frac{5}{25} = \frac{1}{5}$

$\begin{array}{r} .20 = 20\% \\ 5\overline{)1.00} \\ -1.0 \\ \hline 00 \end{array}$

Change each of the following to a percent.

S1. 3 is what percent of 12?

S2. 15 is what percent of 20?

1. 7 is what percent of 28?

2. 20 is what percent of 25?

3. 40 = what percent of 80?

4. 18 is what percent of 20?

5. 12 is what percent of 20?

6. 9 is what percent of 12?

7. 15 = what percent of 20?

8. 24 is what percent of 32?

9. 400 is what percent of 500?

10. 19 is what percent of 20?

1.	
2.	
3.	
4.	
5.	
6.	
7.	
8.	
9.	
10.	
Score	

Problem Solving

A rancher has 800 cows. If he sells 60% of them, how many will he have left?

Review Exercises

1. Write seven and twenty-eight hundredths as a decimal.

2. Write three thousandths as a fraction.

3. Find 40% of 60.

4. Which is the larger decimal? .796 or .9

5. $.15\overline{)15}$

6. 2.08
    x 1.6

Helpful Hints

Use what you have learned to solve the following problems.
1. Write a fraction.
2. Change the fraction to a decimal.
3. Change the decimal to a percent.

Change each of the following to a percent.

S1. 9 is what percent of 12?   S2. 5 is what percent of 25?

1. 2 is what percent of 5?   2. 27 is what percent of 36?

3. 9 is what percent of 10?   4. 9 is what percent of 20?

5. 8 is what percent of 32?   6. 30 is what percent of 40?

7. 60 is what percent of 80?   8. 12 is what percent of 20?

9. 45 is what percent of 50?   10. 13 is what percent of 25?

1.
2.
3.
4.
5.
6.
7.
8.
9.
10.
Score

Problem Solving

A student took a test with 40 questions. If he got 90% of them correct, how many problems did he get correct?

Review Exercises

1. Find 40% of 280.
2. Find 5% of 60.
3. 3 is what % of 5?

4. 15 is what % of 25?
5. Write .471 as a percent.
6. Write $72\frac{601}{100,000}$ as a decimal.

Helpful Hints

When finding the percent first write a fraction, next change the fraction to a decimal, then change the decimal to a percent.

Example:

A team played 20 games and won 15 of them. What percent of the games did they win?

15 is what % of 20?

$$\frac{15}{20} = \frac{3}{4}$$

$$\begin{array}{r} .75 = 75\% \\ 4\overline{)3.00} \\ -28 \\ \hline -20 \\ -20 \\ \hline 0 \end{array}$$

They won 75% of the games.

S1. A test had 25 questions. If Sam got 15 questions correct, what percent did she get correct?

S2. In a class of 20 students, 12 are girls. What percent of the class is girls?

1. On a spelling test with 25 words, Susan got 20 correct. What percent of the words did she get correct?

2. A worker earned 600 dollars. If she put 150 dollars into a savings account, what percent of her earnings did she put into a savings account?

3. A team played 16 games and won 12 of them. What percent did they lose?

4. A quarterback threw 35 passes and 21 were caught. What percent of the passes were caught?

5. $\frac{18}{20}$ of a class was present at school. What percent of the class was present?

6. A class has an enrollment of 30 students. If 24 are present, what percentage absent?

7. A team won 12 games and lost 13 games. What percent of the games played did they win?

8. A school has 300 students. If 120 of them are sixth graders, what percent are sixth graders?

9. On a math test with 50 questions Jill got 49 of them correct. What percent did she get correct?

10. A pitcher threw 12 pitches. If 9 of them were strikes, what percent were strikes?

| 1. |
| 2. |
| 3. |
| 4. |
| 5. |
| 6. |
| 7. |
| 8. |
| 9. |
| 10. |

Problem Solving

A t.v. set costs $420. The sales tax is 8% of the price. What is the total price with tax included?

Score

Review Exercises

1. Change $\frac{11}{20}$ to a percent.

2. Write .7 as a percent.

3. Find 20% of 40.

4. Find 25% of 80.

5. $.12\overline{)60}$

6. $.003\overline{)1.5}$

Helpful Hints

Use what you have learned to solve the following problems.
* Put your answer in a short phrase or sentence.
* If necessary, refer to the example on the previous page.

S1. In a class of 40 students, 10 of them received A's. What percent did not receive A's?

S2. On a spelling test with 12 words, Sean misspelled 3. What percent did he misspell?

1. A team played 15 games and won 12. What percent did they win?

2. A team played 25 games and won 20 of them. What percent of the games did they lose?

3. A class has 30 boys and 20 girls. What percent of the class is boys?

4. A worker earned 500 dollars and deposited 300 dollars of it into a savings account. What percent of his earnings did he deposit?

5. $\frac{9}{15}$ of the students in a school take the bus. What percent of the students take the bus.

6. Santiago has 35 fish. If 14 of them are goldfish, what percent of them are goldfish?

7. A basketball player shot 12 free throws and made 9 of them. What percent of the free throws did he miss?

8. In a class of 50 students, 45 of them have a home computer. What percent of them have a home computer?

9. A quarterback threw 24 passes and 18 were complete. What percent of the passes were completed?

10. In a survey of 40 people it is found that 12 of them have a pet. What percent of those surveyed have a pet?

1.
2.
3.
4.
5.
6.
7.
8.
9.
10.
Score

Problem Solving

Sandy worked five straight days and earned $250.75. What were her average daily earnings?

Review Exercises

1. Find 4% of 80.

2. Find 40% of 80.

3. 12 is what percent of 16?

4. 45 is what percent of 50?

5. 52 - 1.96 =

6. $.06\overline{)12}$

Helpful Hints	To find the whole when the part and the percent are known, simply change the equal sign " = " to the division sign " ÷ ". **Examples:**

6 = 25% of what number?
6 ÷ 25% "Change = to ÷."
6 ÷ .25 "Change % to decimal."

12 = 80% of what number?
12 ÷ 80% "Change = to ÷."
12 ÷ .8 "Change % to decimal."

$$.25\overline{)6.00}\quad\overset{.24}{}$$

*Be careful to move decimal points properly.

$$.8\overline{)12.0}\quad\overset{15.}{}$$

Solve each of the following.

S1. 5 = 25% of what?

S2. 6 is 20% of what?

1. 12 = 25% of what?

2. 32 = 40% of what?

3. 5 is 20% of what?

4. 3 = 75% of what?

5. 12 is 80% of what?

6. 8 = 40% of what?

7. 15 is 25% of what?

8. 15 is 20% of what?

9. 9 is 20% of what?

10. 25 is 20% of what?

1.	
2.	
3.	
4.	
5.	
6.	
7.	
8.	
9.	
10.	
Score	

Problem Solving

Bill took a test with 40 problems and got 36 of them correct. What percent of the problems did he get correct?

Review Exercises

1. Find 12% of 60.

2. Find 30% of 80.

3. 3 is what percent of 12?

4. 6 = what percent of 30?

5. 4 = 25% of what?

6. 5 is 20% of what?

Helpful Hints	Use what you have learned to solve the following problems. Use the following order.
	1. Change = to ÷
	2. Change % to decimal
	3. Divide
	    * Be careful to move decimal points properly.
	    * "Is" means =.

Solve each of the following.

S1. 30 = 15% of what?

S2. 8 is 20% of what?

1. 5 = 25% of what?

2. 20 = 40% of what?

3. 3 is 5% of what?

4. 15 is 30% of what?

5. 10 is 40% of what?

6. 10 = 4% of what?

7. 7 is 20% of what?

8. 9 is 25% of what?

9. 16 = 20% of what?

10. 3 = 2% of what?

1.

2.

3.

4.

5.

6.

7.

8.

9.

10.

Score

Problem Solving	A bakery baked 250 cakes and sold 90% of them. How many cakes were sold?

Review Exercises

1. Find 6% of 200.

2. 12 is what % of 48?

3. 3 = what 20% of what?

4. Change $\frac{12}{15}$ to a decimal.

5. Change 7.009 to a mixed numeral.

6. Change $6\frac{17}{1,000}$ to a decimal.

Helpful Hints

Use what you have learned to solve the following problems.   **Examples:**

5 people got A's on a test. This is 20% of the class. How many people are in the class?

5 = 20% of what number?
5 ÷ 20%
5 ÷ .20

$$2.\overline{\smash{)}5.0.}^{2\,5.}$$

There are 25 in the class.

200 students at a school are 7th graders. If this is 25% of the total students in the school, how many students are there in the school?

200 = 25% of what number?
200 ÷ 25%
200 ÷ .25

$$.25.\overline{\smash{)}200.00.}^{8\,0\,0.}$$

There are 800 students in the school.

S1. A team won 3 games. If this is 20% of the total games played, how many games have they played?

S2. Lucy deposited 150 dollars of her earnings into a savings account. If this was 25% of her earnings, how much did she earn?

1. James has 24 USA stamps in his collection. If that is 20% of his collection, how many stamps are in his collection?

2. A player made 9 shots. This was 75% of her total shots taken. How many shots did the player take.

3. 200 people eat cafeteria food at a school. If this is 40% of the school, how many students are there in the school?

4. 16 = 20% of what?

5. A man spent 8 dollars which was 5% of his earnings. What were his earnings?

6. A farmer sold 25 cows which was 20% of his herd. How many cows were in his herd?

7. 7 is 5% of what?

8. Ted has finished 3 problems on a test. If this is 15% of the problems, how many problems are on the test.

9. 12 players made the team. If this was 15% of all those who tried out, how many tried out for the team?

10. Sophia got 24 problems correct on a test. Her score was 80%. How many problems were on the test?

1.
2.
3.
4.
5.
6.
7.
8.
9.
10.
Score

Problem Solving

Anna bought groceries that cost $63.72. If she paid with four twenty-dollar bills, what is her change?

Review Exercises

1. 7.567 + 85 + .376 =   2. 3.19 - 1.776 =   3.   .616
                                                             x .6

4. 1,000 x 4.5 =   5. .3 $\overline{\smash{)}.027}$ 6. .005 $\overline{\smash{)}16}$

Helpful Hints

Use what you have learned to solve the following problems.
* If necessary, refer to the examples on the previous page.

S1. If you get 35 questions right on a test, and this is 70% of the questions, how many questions are on the test?

S2. 20 people passed a test. This was 16% of those who took the test. How many took the test?

1. There are 3 girls in a class. If this is 20% of the class, how many are in the class?

2. A pitcher threw 9 strikes. If this was 75% of the total pitches thrown, how many pitches were thrown?

3. 25 = 20% of what?

4. There are 8 red marbles in a bag. If this is 40% of all the marbles, how many marbles are in the bag?

5. 15 students in a class were receiving awards. If this is 20% of the class, how many are in the class?

6. 6 is 40% of what?

7. Eva tipped a waiter 6 dollars. This was 15% of the bill. How much was the bill?

8. Mr. Pena paid $6,000 in taxes last year. If this was 25% of his earnings, what were his earnings?

9. Robert has saved 40 dollars. If this is 20% of the cost of a bike that he wants, what is the price of the bike?

10. 4 = 80% of what?

1.

2.

3.

4.

5.

6.

7.

8.

9.

10.

Score

Problem Solving

A man bought a CD player for $15.50. If state tax is 8%, what is the total cost of the CD player with tax included?

Review Exercises

1. $7\overline{)\,.056}$ 2. $1.5\overline{)\,1.35}$ 3. Change $\frac{5}{8}$ to a decimal.

4. 2.13
    $\underline{\text{x }.05}$ 5. $6.5 + 7 + 2.23 =$ 6. 5% of $30 =$

Helpful Hints	Use what you have learned to solve the following problems. **Examples:**

Find 12% of 50.
.12 x 50

   50
$\underline{\text{x }.12}$
 100
 $\underline{\ 50}$
6.00

5 is what percent of 25?

$$\frac{5}{25} = \frac{1}{5}$$

$$5\overline{)\,1.00}^{\,.20} = 20\%$$

6 is 25% of what?
$6 \div 25\%$
$6 \div .25$

$$.25\overline{)\,6.00.}^{\,24.}$$

Solve each of the following.

S1. 4 is what percent of 20?

S2. 3 = 15% of what?

1. Find 20% of 210.

2. Find 6% of 350.

3. 15 is what percent of 60?

4. 5 is 20% of what?

5. 15 = 75% of what?

6. 30% of 200 =

7. 18 is what percent of 24?

8. Find 25% of 64.

9. 3 is 5% of what?

10. 16 is what percent of 80?

Problem Solving	Izzy took a test with 50 problems and got 80% correct. How many problems did he miss?

Answer column:

1.

2.

3.

4.

5.

6.

7.

8.

9.

10.

Score

Review Exercises

1. 120
    x .06

2. 2.45
    x .7

3. $\frac{5}{8}$ x 16 =

4. $5\overline{)1.3}$

5. $8\overline{)2}$

6. $.05\overline{).013}$

Helpful Hints

Use what you have learned to solve the following problems.
* Refer to the examples on previous pages if necessary.

Solve each of the following.

S1. Find 80% of 360.

S2. 7 is 20% of what?

1. 3 is what % of 60?

2. Find 8% of 320.

3. 20 is 25% of what?

4. Find 40% of 60.

5. 12 = 50% of what?

6. 60 = what % of 80?

7. 3 = 50% of what?

8. Find 100% of 320.

9. 4 what % of 20?

10. Find 50% of 60.

1.
2.
3.
4.
5.
6.
7.
8.
9.
10.
Score

Problem Solving

Amy took a spelling test with 12 words on it. If she spelled 9 of the words correctly, what percent of the words did she spell correctly?

Review Exercises

1. Change $\frac{72}{100,000}$ to a decimal.

2. Change 2.0019 to a mixed numeral.

3. Change $\frac{9}{15}$ to a percent.

4. $\frac{3}{5} \times 25 =$

5. $8\overline{).168}$

6. $.3\overline{)2.4}$

Helpful Hints

Use what you have learned to solve the following problems. **Examples:**

A man earns $300 and spends 40% of it. How much does he spend?

Find 40% of 300
.4 x 300

$$\begin{array}{r} 300 \\ \times\ .4 \\ \hline 120 \end{array}$$

He spends $120.

In a class of 25 students 15 are girls. What % are girls?

15 = what % of 25?

$$\frac{15}{25} = \frac{3}{5}$$

.60 = 60%
$5\overline{)3.00}$

60% are girls.

Five students got A's on a test. This is 20% of the class. How many are in the class?

5 = 20% of what?
5 ÷ 20%
5 ÷ .2

$.2\overline{)5.0}$ = 25.

25 are in the class.

S1. On a test with 25 questions, Al got 80% correct. How many questions did he get correct?

S2. A player took 12 shots and made 9. What percent did the player make?

1. A girl spent $5. This was 20% of her earnings. How much were her earnings?

2. Buying an $8,000 car requires a 20% down payment. How much is the down payment?

3. 3 = 10% of what?

4. A team played 20 games and won 18. What % did they lose?

5. A farmer sold 50 cows. If this was 20% of his herd, how many cows were in his herd?

6. 20 = 80% of what?

7. Paul wants a bike that costs $400. If he has saved 60% of this amount, how much has he saved?

8. There are 400 student in a school. 55% are girls. How many boys are there?

9. 12 is what % of 60?

10. Kelly earned 300 dollars and put 70% of it into the bank. How much did she put into the bank?

1.

2.

3.

4.

5.

6.

7.

8.

9.

10.

Score

Problem Solving

Nick's monthly income is $4,800. What is his annual income? (Hint: How many months are in a year?)

Review Exercises

1. 7.68 + 19.7 + 5.364 =

2.   7.123
   - 4.765

3.   3.14
   x 7

4.   .208
   x .06

5. $3\overline{)1.44}$

6. $.15\overline{)1.215}$

Helpful Hints

Use what you have learned to solve the following problems.
* Refer to the examples on the previous page if necessary.

1.
2.
3.
4.
5.
6.
7.
8.
9.
10.
Score

S1. Find 20% of 150.

S2. 6 is 20% of what?

1. 8 is what % of 40?

2. Change $\frac{18}{20}$ to a percent.

3. A school has 600 students. If 5% are absent, how many student are absent?

4. A quarterback threw 24 passes and 75% were caught. How many were caught?

5. Riley has 250 marbles in his collection. If 50 of them are red, what percent of them are red?

6. A team played 60 games and won 45 of them. What % did they win?

7. There are 50 sixth graders in a school. This is 20% of the school. How many students are in the school total?

8. A coat is on sale for $20. This us 80% of the regular price. What is the regular price?

9. Steve has finished $\frac{3}{5}$ of his test. What percent of the test has he finished?

10. Alex wants to buy a computer priced at $640. If sales tax is 8% what is the total cost of the computer?

Problem Solving

Ann took five tests and scored a total of 485 points. What was her average?

197

Final Review of Percents

Change numbers 1 - 5 to a percent.

1. $\dfrac{19}{100} =$
2. $\dfrac{7}{100} =$
3. $\dfrac{9}{10} =$

4. $.27 =$
5. $.3 =$

Change numbers 6 - 8 to a decimal and a fraction reduced to the lowest terms.

6. $3\% = .\underline{\hphantom{00}} = \underline{\hphantom{000}}$
7. $16\% = .\underline{\hphantom{00}} = \underline{\hphantom{000}}$

8. $90\% = .\underline{\hphantom{00}} = \underline{\hphantom{000}}$

Solve the following problems. Label the word problem answers.

9. Find 6% of 280.

10. Find 70% of 450.

11. Find 24% of 400.

12. 3 is what % of 15?

13. 24 = what % of 60?

14. 5 = 20% of what?

15. 6 is 5% of what?

16. Change $\dfrac{27}{36}$ to a percent.

17. Of the 300 students in a school, 40% are girls. How many girls are there in the school?

18. Cloe took a test with 35 questions and got 28 correct. What percent did she get correct?

19. Shawn has finished 18 questions on a test. This is 75% of the test. How many questions are on the test?

20. A team played 65 games and won 80% of them. How many games did the team lose?

1.	
2.	
3.	
4.	
5.	
6.	
7.	
8.	
9.	
10.	
11.	
12.	
13.	
14.	
15.	
16.	
17.	
18.	
19.	
20.	
Score	

Final Test of Percents

Change numbers 1 - 5 to a percent.

1. $\dfrac{7}{10} =$ 2. $\dfrac{7}{100} =$ 3. $\dfrac{1}{10} =$

4. $.05 =$ 5. $.5 =$

Change numbers 6 - 8 to a decimal and a fraction reduced to the lowest terms.

6. $70\% = .$ ___ = ___ 7. $2\% = .$ ___ = ___

8. $15\% = .$ ___ = ___

Solve the following problems, label the word problem answers.

9. Find 40% of 550.

10. Find 25% of 400.

11. Find 3% of 180.

12. 6 = 15% of what?

13. 15 is what % of 50?

14. 20 is 80% of what?

15. Find 8% of 210.

16. 24 is 25% of what?

17. A woman had $4,000 in her savings account. She withdrew 30% of it. How much did she withdraw?

18. On a baseball team there are 6 pitchers. If this is 15% of the team, how many players are on the team?

19. A man must pay a sales tax of 8% when purchasing a car. If the price of the car is $24,000, how much is the sales tax?

20. Rosita had 20 dollars. If she spent 16 dollars, what % of her money did she spend?

1.
2.
3.
4.
5.
6.
7.
8.
9.
10.
11.
12.
13.
14.
15.
16.
17.
18.
19.
20.
Score

Answer Key

Whole Numbers and Integers—Solutions

PAGE 12
Review Exercises:
1. 357
2. 627
3. 13
4. 739
5. 16
6. 49
S1. 819
S2. 933
1. 113
2. 783
3. 1,000
4. 1,072
5. 1,497
6. 687
7. 1,168
8. 120
9. 1,673
10. 412
Problem Solving: 134 students

PAGE 13
Review Exercises:
1. 449
2. 731
3. 83
4. 1,093
5. 1,178
6. 944
S1. 1,285
S2. 734
1. 1,264
2. 135
3. 1,767
4. 675
5. 1,913
6. 1,551
7. 329
8. 1,180
9. 288
10. 1,578
Problem Solving: $220

PAGE 14
Review Exercises:
1. 191
2. 1,225
3. 2,080
4. 1,906
5. 137
6. 62
S1. 7,619
S2. 44,027
1. 10,395
2. 24,708
3. 3,628,991
4. 7,497,988
5. 71,078
6. 94,180
7. 7,697
8. 105,598
9. 118,112
10. 899,935
Problem Solving: 36,283,995

PAGE 15
Review Exercises:
1. 885
2. 53,203
3. 134,680
4. 33,104,468
5. 211,628
6. 10,562,592
S1. 100,463
S2. 6,301,851
1. 78,734
2. 8,976,984
3. 12,915
4. 86,835
5. 20,422
6. 1,166,094
7. 212,299
8. 14,362,605
9. 97,625
10. 42,865
Problem Solving: 8,044 students

PAGE 16
Review Exercises:
1. 154
2. 43,939
3. 19,731,174
4. 1,615
5. 1,370,727
6. 337
S1. 384
S2. 2,181
1. 315
2. 571
3. 5,395
4. 2,775
5. 648
6. 5,098
7. 716
8. 3,389
9. 6,470
10. 491
Problem Solving: 97 students

PAGE 17
Review Exercises:
1. 1,883
2. 278
3. 6,057
4. 2,126
5. 4,861
6. 41,989,515
S1. 559
S2. 7,366
1. 473
2. 175
3. 1,885
4. 4,159
5. 4,667
6. 6,279
7. 11,874
8. 1,517
9. 2,369
10. 2,928
Problem Solving: $58

PAGE 18
Review Exercises:
1. 1,199
2. 6,766
3. 3,451
4. 8,692
5. 1,263
6. 2,656
S1. 245
S2. 433
1. 23
2. 529
3. 2,667
4. 462
5. 7,242
6. 3,668
7. 1,337
8. 2,216
9. 6,678
10. 7,485
Problem Solving: 291 seats

PAGE 19
Review Exercises:
1. 224
2. 4,214
3. 277
4. 4,249
5. 3,979
6. 4,232
S1. 4,546
S2. 6,033
1. 823
2. 2,428
3. 664
4. 4,803
5. 1,649
6. 42,458
7. 68,296
8. 4,494
9. 443
10. 47,192
Problem Solving: $16,500

PAGE 20
Review Exercises:
1. 830
2. 51,656
3. 1,969
4. 4,691
5. 4,236
6. 1,827
S1. 24,907
S2. 2,674
1. 632
2. 461
3. 13,049
4. 2,211
5. 4,303
6. 165
7. 9,979
8. 992
9. 778
10. 152,371
Problem Solving: $95

PAGE 21
Review Exercises:
1. 44
2. 73,212
3. 1,251
4. 64,738
5. 2,474
6. 79,922
S1. 5,655
S2. 52,811
1. 676
2. 5,764
3. 10,808
4. 27,891
5. 198
6. 344
7. 1,793
8. 67,384
9. 4,509
10. 37,046
Problem Solving: 8603 people

PAGE 22
Review Exercises:
1. 234
2. 1,052
3. 2,335
4. 148
5. 7,216
6. 40,117
S1. 1,578
S2. 19,524
1. 201
2. 444
3. 2,562
4. 25,284
5. 24,448
6. 43,536
7. 49,252
8. 15,396
9. 52,572
10. 34,848
Problem Solving: 2,555 days

PAGE 23
Review Exercises:
1. 1,372
2. 21,448
3. 2,803
4. 4,765
5. 874
6. 3,052
S1. 43,404
S2. 72,056
1. 4,368
2. 81,108
3. 166,082
4. 45,792
5. 87,570
6. 28,036
7. 21,144
8. 56,084
9. 142,368
10. 74,574
Problem Solving: 152 points

Whole Numbers and Integers—Solutions

PAGE 24
Review Exercises:
1. 4,350
2. 27,108
3. 15,846
4. 2,010
5. 3,327
6. 1,462
S1. 1,032
S2. 13,038
1. 4,346
2. 782
3. 3,995
4. 10,900
5. 16,800
6. 33,182
7. 592
8. 4,324
9. 34,335
10. 15,180
Problem Solving: 875 desks

PAGE 25
Review Exercises:
1. 2,839
2. 22,015
3. 1,728
4. 10,575
5. 15,726
6. 5,563
S1. 10,452
S2. 15,664
1. 1,300
2. 1,288
3. 2,530
4. 13,824
5. 18,848
6. 35,700
7. 9,476
8. 12,168
9. 4,550
10. 44,384
Problem Solving: 1200 girls

PAGE 26
Review Exercises:
1. 1,512
2. 11,396
3. 14,000
4. 6,255
5. 154
6. 4,164
S1. 64,719
S2. 380,824
1. 77,805
2. 33,702
3. 73,688
4. 183,092
5. 224,802
6. 107,600
7. 175,577
8. 67,808
9. 411,000
10. 162,504
Problem Solving: 158,600 cars

PAGE 27
Review Exercises:
1. 11,853
2. 824
3. 13,032
4. 1,058
5. 7,890
6. 370,304
S1. 99,900
S2. 317,100
1. 649,602
2. 371,500
3. 140,505
4. 489,240
5. 662,592
6. 229,272
7. 88,914
8. 427,000
9. 692,804
10. 612,381
Problem Solving: 784 students

PAGE 28
Review Exercises:
1. 2,448
2. 864
3. 25,334
4. 777
5. 4,452
6. 11,016
S1. 10,650
S2. 260,525
1. 185
2. 4,230
3. 16,422
4. 4,176
5. 15,435
6. 70,016
7. 10,890
8. 459,900
9. 20,130
10. 125,356
Problem Solving: 10,080 minutes

PAGE 29
Review Exercises:
1. 726
2. 3,135
3. 6,329
4. 1,498
5. 15,862
6. 182,000
S1. 30,035
S2. 19,085
1. 2,020
2. 42,012
3. 62,412
4. 7,680
5. 45,234
6. 307,824
7. 2,160
8. 70,704
9. 47,905
10. 158,704
Problem Solving: 101 people

Whole Numbers and Integers—Solutions

PAGE 30
Review Exercises:
1. 307
2. 1,735
3. 88,625
4. 6,917
5. 2,992
6. 61,800
S1. 4 r1
S2. 9 r4
1. 24 r2
2. 5 r7
3. 13
4. 18 r3
5. 12
6. 15 r4
7. 9 r5
8. 24
9. 16 r1
10. 23 r1
Problem Solving: 18 pencils

PAGE 31
Review Exercises:
1. 8 r1
2. 17 r4
3. 6 r3
4. 106
5. 1155
6. 1551
S1. 18 r1
S2. 4 r3
1. 9 r2
2. 15
3. 11 r5
4. 48 r1
5. 8 r4
6. 5 r4
7. 15 r2
8. 24 r3
9. 8 r1
10. 13
Problem Solving: $78,000

PAGE 32
Review Exercises:
1. 6 r2
2. 13 r4
3. 4,486
4. 3,682
5. 3,772
6. 222
S1. 261 r1
S2. 45 r4
1. 324
2. 209 r1
3. 133 r2
4. 121 r1
5. 114 r4
6. 141 r4
7. 312
8. 121 r2
9. 258 r2
10. 89 r5
Problem Solving: 36 packages

PAGE 33
Review Exercises:
1. 19
2. 12 r5
3. 123 r1
4. 74 r4
5. 132
6. 34 r2
S1. 187 r1
S2. 94 r1
1. 114 r4
2. 78 r6
3. 97
4. 122
5. 127
6. 41 r2
7. 133 r1
8. 99
9. 155 r3
10. 101 r3
Problem Solving: 5,400 miles

PAGE 34
Review Exercises:
1. 23
2. 133 r2
3. 10,011
4. 5,373,870
5. 552,600
6. 773
S1. 3017 r1
S2. 670 r2
1. 4,617 r1
2. 3,681
3. 1,136 r5
4. 448
5. 1,533 r1
6. 1,780 r3
7. 12,247
8. 12,181
9. 10,309
10. 4,402 r4
Problem Solving: $3,310

PAGE 35
Review Exercises:
1. 175 r1
2. 201 r2
3. 376 r2
4. 1,235
5. 2,208 r4
6. 6,332
S1. 662 r4
S2. 2,795 r1
1. 1,374 r1
2. 345 r1
3. 1,112 r6
4. 887 r2
5. 1551 r3
6. 1,002
7. 5,574 r1
8. 3,186 r3
9. 15,263 r1
10. 4,714 r4
Problem Solving: 44 miles

Whole Numbers and Integers—Solutions

PAGE 36
Review Exercises:
1. 782
2. 25,368
3. 11,102
4. 125
5. 193 r5
6. 119,507
S1. 300 r1
S2. 1,080 r3
1. 80 r2
2. 1,700
3. 2,103 r1
4. 445 r6
5. 254 r2
6. 1,200
7. 2,002 r1
8. 973 r8
9. 904 r1
10. 804 r1
Problem Solving: 12,500 sheets

PAGE 37
Review Exercises:
1. 2,459
2. 319,774
3. 85 r5
4. 1,000 r3
5. 1,525 r1
6. 281 r2
S1. 123
S2. 1,462
1. 61 r2
2. 17 r4
3. 202 r1
4. 1,001 r5
5. 1,622 r4
6. 349 r2
7. 1,677 r7
8. 2,877 r2
9. 11,872 r2
10. 3,071
Problem Solving: 556 containers

PAGE 38
Review Exercises:
1. 119 r4
2. 1,142
3. 226,800
4. 1,001 r2
5. 233 r1
6. 16,253
S1. 5 r26
S2. 143 r12
1. 3 r28
2. 7 r26
3. 8 r38
4. 11 r59
5. 56 r2
6. 69 r16
7. 224 r12
8. 101 r6
9. 94 r26
10. 59 r36
Problem Solving: 256 boxes

PAGE 39
Review Exercises:
1. 34 r1
2. 1,002 r5
3. 1,577
4. 10,944
5. 5,132
6. 2,041
S1. 2 r37
S2. 126 r52
1. 19 r16
2. 9 r7
3. 77 r2
4. 246 r16
5. 81 r54
6. 142 r48
7. 199 r26
8. 44 r6
9. 27 r49
10. 241 r26
Problem Solving: 768 students

PAGE 40
Review Exercises:
1. 1,046
2. 5,768
3. 250
4. 12 r9
5. 5 r17
6. 131 r36
S1. 21 r1
S2. 8 r23
1. 22 r34
2. 22 r1
3. 12 r1
4. 9 r15
5. 24 r10
6. 31 r15
7. 12 r15
8. 11 r26
9. 44 r12
10. 20 r26
Problem Solving: 14 gallons

PAGE 41
Review Exercises:
1. 11,410
2. 162,000
3. 206
4. 2,244
5. 16 r26
6. 81 r82
S1. 22 r10
S2. 7 r10
1. 13 r30
2. 22 r18
3. 13 r2
4. 36 r4
5. 32 r9
6. 24 r18
7. 10 r46
8. 31 r2
9. 34 r24
10. 17 r36
Problem Solving: 185 seats

PAGE 42
Review Exercises:
1. 3 r8
2. 13 r13
3. 8 r3
4. 13 r10
5. 8 r19
6. 14 r13
S1. 3 r71
S2. 19 r4
1. 5 r7
2. 5 r82
3. 54 r15
4. 8 r2
5. 5 r3
6. 63 r4
7. 19 r3
8. 19 r31
9. 20 r1
10. 43 r4
Problem Solving: 900 parts

PAGE 43
Review Exercises:
1. 78
2. 17,332
3. 1,472
4. 5,175
5. 250,800
6. 388,367
S1. 29 r21
S2. 37 r16
1. 31 r14
2. 32 r5
3. 30 r2
4. 20 r16
5. 29 r28
6. 49
7. 43 r13
8. 19 r31
9. 63 r4
10. 8 r10
Problem Solving: $53.00

PAGE 44
Review Exercises:
1. 365
2. 467
3. 30,658
4. 1,765
5. 2,969
6. 1,347
S1. 215 r5
S2. 207 r11
1. 125 r20
2. 148 r15
3. 214 r17
4. 165 r11
5. 298 r10
6. 306 r1
7. 32 r62
8. 80 r5
9. 142
10. 750 r6
Problem Solving: 440 miles

PAGE 45
Review Exercises:
1. 35 r1
2. 872 r1
3. 107 r42
4. 42 r27
5. 12 r2
6. 8 r33
S1. 48 r5
S2. 212 r11
1. 94 r22
2. 77 r35
3. 114 r14
4. 94 r28
5. 116 r5
6. 51 r54
7. 317 r17
8. 163 r19
9. 401 r1
10. 43 r3
Problem Solving:
11 cartons, 27 left

PAGE 46
Review Exercises:
1. 107,083
2. 4,329
3. 155,935
4. 218,255
5. 5,110
6. 7,392
S1. 95 r21
S2. 55 r 21
1. 38 r1
2. 331 r3
3. 279 r2
4. 16 r7
5. 7 r76
6. 41 r72
7. 25 r16
8. 12 r2
9. 54 r50
10. 253
Problem Solving: 51 students

PAGE 47
Review Exercises:
1. 34
2. 654 r2
3. 9 r9
4. 45 r17
5. 11 r25
6. 73 r11
S1. 96 r16
S2. 77 r17
1. 19 r2
2. 325 r5
3. 778
4. 16 r7
5. 6 r36
6. 84 r56
7. 5 r16
8. 12 r50
9. 23 r48
10. 130 r28
Problem Solving: $960

Whole Numbers and Integers—Solutions

PAGE 48
1. 1,109
2. 2,899
3. 8,1387
4. 10,675
5. 8,474
6. 345
7. 2,815
8. 6,285
9. 4,012
10. 6,285
11. 261
12. 45,794
13. 3,087
14. 61,484
15. 113,702
16. 141 r2
17. 282 r5
18. 21 r27
19. 150 r17
20. 44 r26

PAGE 49
1. 1,846
2. 89,844
3. 32,904
4. 8,638
5. 26,297
6. 435
7. 8,215
8. 44,499
9. 2,643
10. 55,807
11. 2,156
12. 63,090
13. 2,523
14. 38,142
15. 771,363
16. 58 r6
17. 232
18. 5 r39
19. 142 r28
20. 594 r10

PAGE 50
Review Exercises:
1. 1,840
2. 6,894
3. 3,783
4. 233 r6
5. 5 r17
6. 17 r39
S1. 6
S2. -23
1. 9
2. -18
3. -16
4. -27
5. 7
6. 38
7. -180
8. -29
9. -70
10. -20
Problem Solving: 6 buses

PAGE 51
Review Exercises:
1. -14
2. 5
3. -78
4. 148,131
5. 12 r25
6. 6 r5
S1. 9
S2. -185
1. -9
2. -936
3. -37
4. -141
5. 37
6. -215
7. -901
8. -99
9. -641
10. 154
Problem Solving: $31

PAGE 52
Review Exercises:
1. 21,063
2. 1,257
3. 58,398
4. -15
5. -188
6. 37
S1. -2
S2. -5
1. 3
2. -3
3. -6
4. 13
5. -6
6. -8
7. -33
8. -9
9. -26
10. -89
Problem Solving: 89

PAGE 53
Review Exercises:
1. 24 r7
2. 9 r24
3. 21 r25
4. 62 r3
5. 2
6. -72
S1. -6
S2. 9
1. -7
2. -22
3. -55
4. 14
5. -34
6. 14
7. -24
8. -108
9. 33
10. -57
Problem Solving: 24 packages

PAGE 54
Review Exercises:
1. -28
2. 8
3. -15
4. 241
5. 1,229
6. 17,292
S1. -13
S2. 1
1. 15
2. -3
3. 8
4. -31
5. 66
6. 34
7. -20
8. -18
9. -93
10. 31
Problem Solving: -8°

PAGE 55
Review Exercises:
1. -9
2. -10
3. -30
4. -42
5. 20
6. -16
S1. -28
S2. 112
1. 38
2. -8
3. 12
4. -46
5. 71
6. -79
7. -14
8. 48
9. -434
10. -16
Problem Solving: $195

PAGE 56
Review Exercises:
1. -16
2. -2
3. 40
4. 21
5. -4
6. -107
S1. -46
S2. 22
1. -105
2. 15
3. -21
4. 60
5. -23
6. 9
7. -29
8. 19
9. 76
10. -924
Problem Solving: -34

PAGE 57
Review Exercises:
1. -185
2. -171
3. 361
4. 156
5. -1,014
6. -22
S1. -965
S2. 157
1. -65
2. 72
3. 71
4. -53
5. 41
6. -95
7. -119
8. -25
9. -68
10. -738
Problem Solving: 21°

PAGE 58
Review Exercises:
1. 766
2. 749
3. 8,208
4. 27 r23
5. -64
6. 10
S1. 48
S2. -136
1. 174
2. -80
3. 504
4. -594
5. -1440
6. -171
7. 56
8. 384
9. -272
10. 646
Problem Solving: $330

PAGE 59
Review Exercises:
1. 42
2. -156
3. -300
4. -12
5. 29
6. 15
S1. -180
S2. 320
1. 110
2. -110
3. 432
4. -420
5. -1,600
6. 375
7. -1,320
8. 720
9. -576
10. 2,400
Problem Solving: 105 feet

Whole Numbers and Integers—Solutions

PAGE 60
Review Exercises:
1. -7
2. 2
3. 42
4. 252
5. -23
6. 36
S1. 64
S2. -30
1. -60
2. -126
3. 360
4. 180
5. -12
6. -120
7. -48
8. 192
9. 252
10. 440
Problem Solving: 3,680 miles

PAGE 61
Review Exercises:
1. 24
2. -105
3. 44
4. -84
5. -36
6. -24
S1. 378
S2. -180
1. 120
2. -105
3. -144
4. 450
5. -200
6. 120
7. -336
8. 360
9. 252
10. 216
Problem Solving: 375 cows

PAGE 62
Review Exercises:
1. -1
2. 2
3. 5
4. -75
5. 18
6. 60
S1. -3
S2. 15
1. -18
2. -48
3. 21
4. -51
5. -26
6. -48
7. -8
8. -17
9. 13
10. 19
Problem Solving: 12 packs

PAGE 63
Review Exercises:
1. 3 r2
2. 199 r3
3. 14 r56
4. 6 r34
5. 5 r26
6. 15 r9
S1. -32
S2. 6
1. -18
2. -81
3. -23
4. 707
5. -72
6. 22
7. 16
8. -5
9. 26
10. 667
Problem Solving:
    59 miles per hour

PAGE 64
Review Exercises:
1. -96
2. 12
3. 24
4. -50
5. -4
6. 6
S1. -1
S2. -12
1. 2
2. 3
3. -4
4. -18
5. -3
6. 2
7. 1
8. -3
9. -3
10. 2
Problem Solving: $1.75

PAGE 65
Review Exercises:
1. 21
2. 25,368
3. 204
4. -17
5. 12
6. -2
S1. 2
S2. -10
1. -4
2. -8
3. -6
4. -10
5. 1
6. 2
7. -2
8. -4
9. -1
10. -2
Problem Solving: 95

Whole Numbers and Integers—Solutions

PAGE 66

Review Exercises:

1. 311
2. 8,897
3. 63,448
4. 56,608
5. 6,576
6. 4,337

S1. 360
S2. 2

1. -90
2. -120
3. 24
4. 15
5. -8
6. 56
7. 36
8. -1
9. -12
10. 1

Problem Solving: 90

PAGE 67

Review Exercises:

1. 25 r1
2. 250
3. 8 r33
4. 18 r27
5. 12 r13
6. 75 r6

S1. 144
S2. -5

1. -150
2. 90
3. 80
4. 25
5. -9
6. 37
7. -2
8. -1
9. -2
10. -3

Problem Solving: 11,234

PAGE 68

1. -3
2. 3
3. -13
4. 3
5. -26
6. -2
7. 12
8. -10
9. -28
10. -1
11. -60
12. 81
13. 72
14. 120
15. -3
16. 78
17. 32
18. -2
19. -8
20. 10

PAGE 69

1. 6
2. -6
3. -18
4. 6
5. -2
6. -3
7. 31
8. -51
9. -49
10. -64
11. -112
12. 150
13. 144
14. 90
15. 9
16. 138
17. -221
18. 2
19. -9
20. -1

PAGE 70

1. 753
2. 1,918
3. 23,407
4. 10,882
5. 8,633
6. 488
7. 3,624
8. 2,764
9. 1,704
10. 6,009
11. 348
12. 28,528
13. 1,620
14. 18,424
15. 183,521
16. 156 r2
17. 357
18. 14 r8
19. 289 r2
20. 74 r11

PAGE 71

1. 2
2. -2
3. -12
4. 0
5. -34
6. -2
7. 11
8. -19
9. -28
10. -5
11. -32
12. 228
13. 120
14. 42
15. -7
16. 112
17. 17
18. -4
19. -8
20. -8

PAGE 74
Review Exercises:
1. 72
2. 549
3. 1,264
4. 328
5. 448
6. 324
S1. 1/4; 3/4
S2. 5/6; 1/6
1. 6/8, 3/4; 2/8, 1/4
2. 1/3; 2/3
3. 2/4, 1/2; 2/4, 1/2
4. 3/8, 5/8
5. 5/8; 3/8
6. 2/3; 1/3
7. 5/6; 1/6
8. 1/8; 7/8
9. 1/3, 2/6; 2/3, 4/6
10. 7/8; 1/8
Problem Solving: 360 crayons

PAGE 75
Review Exercises:
1. 127
2. 629
3. 536
4. 1,498
5. 435
6. 173
S1. 1/3; 2/3
S2. 4/8, 2/4, 1/2; 4/8, 2/4, 1/2
1. 7/8; 1/8
2. 2/8, 1/4; 6/8, 3/4
3. 6/8, 3/4; 2/8, 1/4
4. 1/3; 2/3
5. 4/8, 2/4, 1/2; 4/8, 2/4, 1/2
6. 3/4; 1/4
7. 3/9, 1/3; 6/9, 2/3
8. 6/9, 2/3; 3/9, 1/3
9. 3/8; 5/8
10. 2/4, 1/2; 2/4, 1/2
Problem Solving: $771

PAGE 76
Review Exercises:
1. 1/2
2. 6/9, 2/3
3. 3/9, 1/3
4. 3/6, 1/2
5. 2/8, 1/4
6. 6/8, 3/4
S1. answers vary, one fifth
S2. answers vary, three eighths
1. answers vary, three fourths
2. answers vary, two ninths
3. answers vary, four fifths
4. answers vary, four sevenths
5. answers vary, seven twelfths
6. answers vary, seven eighths
7. answers vary, four eighths
8. answers vary, two thirds
9. answers vary, five sixths
10. answers vary, four sixths
Problem Solving: 440 miles

PAGE 77
Review Exercises:
1. 457
2. answers vary
3. 6/8, 3/4
4. 465
5. 682
6. 192
S1. answers vary, one third
S2. answers vary, three tenths
1. answers vary, five eighths
2. answers vary, one sixth
3. answers vary, one eighth
4. answers vary, three twelfths
5. answers vary, two thirds
6. answers vary, seven ninths
7. answers vary, two sevenths
8. answers vary, six ninths
9. answers vary, nine tenths
10. answers vary, two fourths
Problem Solving: 27

PAGE 78
Review Exercises:
1. 2/4, 1/2
2. 6/8, 3/4
3. answers vary
4. 792
5. 437
6. 367
S1. 1/2
S2. 3/4
1. 4/5
2. 3/4
3. 1/2
4. 4/5
5. 2/3
6. 2/3
7. 3/5
8. 5/8
9. 5/6
10. 3/4
Problem Solving: 12 gallons

PAGE 79
Review Exercises:
1. 3/4
2. 4/5
3. 89
4. answers vary
5. 1,363
6. 13,820
S1. 3/4
S2. 5/6
1. 3/5
2. 3/4
3. 8/15
4. 7/8
5. 3/4
6. 5/8
7. 15/16
8. 5/6
9. 8/9
10. 5/6
Problem Solving: 14 miles

PAGE 80

Review Exercises:

1. 348
2. 789
3. 6,237
4. 3/4
5. 457
6. 5/6
S1. 15
S2. 9
1. 10
2. 12
3. 18
4. 22
5. 35
6. 18
7. 6
8. 40
9. 16
10. 33

Problem Solving: $30

PAGE 81

Review Exercises:

1. 3/4
2. 281
3. 4/6, 2/3
4. 171
5. 21
6. 7,285
S1. 18
S2. 30
1. 12
2. 44
3. 35
4. 30
5. 18
6. 9
7. 6
8. 33
9. 32
10. 49

Problem Solving: 39 miles per gallon

PAGE 82

Review Exercises:

1. 5/8
2. 2/3
3. 14/25
4. 235
5. 1981
6. 1/2, 2/4, 4/8
S1. 1 1/4
S2. 1 1/2
1. 2 3/4
2. 1 4/7
3. 3
4. 5 1/2
5. 3 1/12
6. 4
7. 5 1/2
8. 6 2/3
9. 2 1/4
10. 5 2/5

Problem Solving: 3,734

PAGE 83

Review Exercises:

1. 1 4/7
2. 9
3. 5/6
4. 6/9, 2/3
5. 6,262
6. 1,980
S1. 1 4/15
S2. 3 1/4
1. 2 1/2
2. 1 1/8
3. 5 2/5
4. 5 1/3
5. 6 1/4
6. 2 1/2
7. 8 1/2
8. 1 7/9
9. 6 2/5
10. 1 2/3

Problem Solving: $42

PAGE 84

Review Exercises:

1. 1 4/5
2. 4/5
3. 2 2/9
4. 21
5. 2/8, 1/4
6. 8
S1. 5/2
S2. 19/3
1. 36/5
2. 11/3
3. 21/4
4. 19/4
5. 17/6
6. 57/8
7. 23/3
8. 35/4
9. 37/4
10. 15/8

Problem Solving: 18 gallons

PAGE 85

Review Exercises:

1. 33
2. 9
3. 2 1/7
4. 4/5
5. answers vary
6. 2/3
S1. 17/2
S2. 19/4
1. 15/2
2. 17/3
3. 43/8
4. 8/3
5. 28/3
6. 38/5
7. 23/3
8. 33/2
9. 23/5
10. 23/4

Problem Solving: 6 packages

PAGE 86
Review Exercises:
1. 1 3/7
2. 23/5
3. 5/7
4. 21/4
5. 1 1/3
6. answers vary
S1. 4/5
S2. 1 1/4
1. 7/9
2. 3/4
3. 1 1/2
4. 9/10
5. 1 1/4
6. 1 1/4
7. 1 4/5
8. 1 1/4
9. 1 3/5
10. 1 2/3
Problem Solving: 1 1/4 cups

PAGE 87
Review Exercises:
1. 11/2
2. 3 1/5
3. 4/5
4. 1 1/2
5. answers vary
6. 9/10
S1. 4/5
S2. 1 1/2
1. 3/4
2. 7/10
3. 1 2/5
4. 1 1/8
5. 1 1/12
6. 1 3/4
7. 1 2/3
8. 1 3/13
9. 2
10. 1 1/5
Problem Solving: 1 3/5 inches

PAGE 88
Review Exercises:
1. 1 1/8
2. 3/5
3. 8/9
4. 5 1/4
5. 15/4
6. 1
S1. 6 4/5
S2. 8 1/4
1. 7 6/7
2. 6 1/5
3. 12
4. 6 1/3
5. 7 2/7
6. 7 1/2
7. 10 1/5
8. 6 2/15
9. 10
10. 12 1/5
Problem Solving: 1 3/4 cups

PAGE 89
Review Exercises:
1. 2/3
2. 4/5
3. 1 1/5
4. 15/16
5. 23/6
6. 5 2/3
S1. 7 4/5
S2. 9 3/5
1. 5 4/5
2. 8 2/5
3. 15
4. 13 1/2
5. 30 1/5
6. 17 1/3
7. 8 1/15
8. 5 1/2
9. 11
10. 13 5/11
Problem Solving: 1/2 mile

PAGE 90
Review Exercises:
1. 1 1/2
2. 4/5
3. 1 2/5
4. 1 1/4
5. 4/5
6. 9 4/7
S1. 1 1/4
S2. 8 1/5
1. 2/3
2. 8
3. 1 3/5
4. 9 1/3
5. 11 1/2
6. 1 2/3
7. 5/8
8. 13
9. 1 1/4
10. 6 1/3
Problem Solving: 14 1/2 hours

PAGE 91
Review Exercises:
1. answers vary
2. 3 1/2
3. 7/10
4. 1 1/3
5. 1
6. 16 1/2
S1. 7 6/7
S2. 1 1/2
1. 1/2
2. 9 1/4
3. 1 7/8
4. 13 1/3
5. 4/5
6. 10 1/5
7. 1 1/6
8. 13 2/3
9. 11 7/8
10. 1 1/9
Problem Solving: 90

PAGE 92
Review Exercises:
1. 1/2
2. 1 2/7
3. 1 1/4
4. 8
5. 15 4/5
6. 10 1/2
S1. 1/2
S2. 5/8
1. 3/5
2. 3/5
3. 3/4
4. 3/4
5. 7/9
6. 1/2
7. 4/5
8. 2/7
9. 3/4
10. 1/5
Problem Solving: 1/2 pounds

PAGE 93
Review Exercises:
1. 4/5
2. 9 2/5
3. 23/3
4. 4/5
5. 1 2/5
6. 2 5/7
S1. 1/2
S2. 7/8
1. 3/4
2. 4/5
3. 1/4
4. 1/3
5. 3/4
6. 1/3
7. 1/3
8. 3/5
9. 4/5
10. 1/4
Problem Solving: 10 3/4 hours

PAGE 94
Review Exercises:
1. 1/2
2. 10
3. 1/4
4. 8
5. 26 1/3
6. 1 5/8
S1. 3 1/2
S2. 2 3/5
1. 4 1/4
2. 4 1/2
3. 3 5/8
4. 5 3/4
5. 2 3/5
6. 2 3/5
7. 3
8. 1 2/3
9. 4 5/7
10. 5 3/5
Problem Solving: 11 1/2 miles

PAGE 95
Review Exercises:
1. 1 4/15
2. 3/4
3. 15/16
4. 1 1/7
5. 3/15, 1/5
6. 8 1/4
S1. 1 1/2
S2. 5 1/2
1. 6 1/3
2. 5 2/3
3. 4 3/4
4. 3 1/2
5. 2 8/11
6. 4 5/6
7. 6 1/2
8. 4 3/4
9. 4 7/10
10. 3 1/3
Problem Solving: 4 Gallons, $12

PAGE 96
Review Exercises:
1. 3/4
2. 2 3/4
3. 3 3/4
4. 1 1/2
5. 6 4/5
6. 9 2/15
S1. 3 2/5
S2. 8 3/8
1. 4 7/8
2. 2 4/7
3. 11 8/15
4. 4 1/10
5. 12 7/8
6. 12 6/7
7. 11 2/11
8. 5 4/5
9. 23 4/7
10. 15 8/15
Problem Solving: 9 1/4 yards

PAGE 97
Review Exercises:
1. answers vary
2. 4 2/3
3. 5 3/4
4. 1 3/4
5. 3/5
6. 19/5
S1. 8 1/8
S2. 9 4/7
1. 4 3/4
2. 32 5/8
3. 3 1/16
4. 8 3/8
5. 10 23/24
6. 15 2/5
7. 39 11/24
8. 8 11/24
9. 12 1/20
10. 24 1/5
Problem Solving: 11 1/2 gallons

Fractions—Solutions

PAGE 98

Review Exercises:
1. 4 2/3
2. 9 2/15
3. 1 6/25
4. 3/5
5. 4 1/4
6. 1 1/6
S1. 6 1/5
S2. 3 3/5
1. 1 2/7
2. 2/3
3. 11 2/9
4. 6 3/4
5. 9 3/4
6. 6 1/2
7. 8 1/3
8. 4 2/3
9. 7 3/5
10. 1 9/10
Problem Solving: 450 miles per hour

PAGE 99

Review Exercises:
1. 2 1/2
2. 6 4/15
3. 17 2/5
4. 1 1/2
5. 5/8
6. 21/4
S1. 4 3/4
S2. 8 1/2
1. 2/3
2. 1 4/9
3. 11 1/7
4. 6 2/5
5. 68 1/4
6. 3 1/4
7. 14 1/5
8. 4 1/2
9. 11
10. 2
Problem Solving: 79 desks

PAGE 100

Review Exercises:
1. 1 1/4
2. 1 1/2
3. 7/8
4. 17/5
5. 1 2/3
6. 4 1/4
S1. 20
S2. 24
1. 18
2. 12
3. 21
4. 24
5. 36
6. 56
7. 48
8. 60
9. 48
10. 48
Problem Solving: 26 students

PAGE 101

Review Exercises:
1. 1 1/5
2. 16 1/4
3. 9 7/8
4. 2/5
5. 6 3/8
6. 5 1/3
S1. 24
S2. 12
1. 18
2. 28
3. 30
4. 90
5. 36
6. 36
7. 12
8. 48
9. 72
10. 30
Problem Solving: 3 1/4 gallons

PAGE 102

Review Exercises:
1. 1 1/3
2. 1/4
3. 30
4. 4/5
5. 5 2/11
6. 7 1/2
S1. 7/12
S2. 1 3/10
1. 8/9
2. 1 1/6
3. 11/12
4. 1 5/12
5. 1 1/4
6. 1 1/4
7. 11/12
8. 1 1/36
9. 1 3/22
10. 13/24
Problem Solving: 1 1/2 hours

PAGE 103

Review Exercises:
1. 1 1/5
2. 8 1/4
3. 4 3/10
4. 2 5/8
5. 4 1/2
6. 24
S1. 7/9
S2. 1 7/12
1. 11/12
2. 1 3/20
3. 22/25
4. 1 5/12
5. 1 1/15
6. 1 4/15
7. 1 13/24
8. 14/33
9. 19/24
10. 23/30
Problem Solving: $36

PAGE 104
Review Exercises:
1. 3/5
2. 1
3. 10 2/5
4. 11/12
5. 1 3/20
6. 1 1/4
S1. 2/9
S2. 7/12
1. 3/40
2. 17/30
3. 19/30
4. 1/4
5. 1/6
6. 7/18
7. 1/4
8. 3/10
9. 33/56
10. 2/5
Problem Solving: 4 3/4 pounds

PAGE 105
Review Exercises:
1. 3/5
2. 3/4
3. 1 2/3
4. 4 1/12
5. 8 1/4
6. 19/30
S1. 1/4
S2. 27/40
1. 1/2
2. 9/16
3. 3/8
4. 3/4
5. 5/12
6. 13/18
7. 3/10
8. 2/5
9. 1/12
10. 17/24
Problem Solving: 7 boxes, 9 left over

PAGE 106
Review Exercises:
1. 1 3/4
2. 1 1/15
3. 13/15
4. 1/4
5. 4 2/5
6. 2/15
S1. 5/8
S2. 1 3/8
1. 8/9
2. 1/12
3. 3/5
4. 1 3/10
5. 13/22
6. 1/2
7. 1/2
8. 21/25
9. 5/14
10. 17/20
Problem Solving: 1 5/12 pounds

PAGE 107
Review Exercises:
1. 8 1/2
2. 12 2/5
3. 10 3/5
4. 4 1/2
5. 3 2/5
6. 8 1/5
S1. 1/2
S2. 1 1/12
1. 3/4
2. 3/4
3. 13/30
4. 1/16
5. 17/24
6. 11/15
7. 1/8
8. 7/8
9. 7/10
10. 1 2/9
Problem Solving: 4 1/4 hours

PAGE 108
Review Exercises:
1. 23/7
2. 1 2/15
3. 4/7
4. answers vary
5. 1 1/8
6. 1 1/8
S1. 7 11/15
S2. 8 1/12
1. 5 1/4
2. 7 5/6
3. 7 9/10
4. 8 1/5
5. 11 2/9
6. 7 1/10
7. 8 3/14
8. 11 1/18
9. 7 10/21
10. 11 5/24
Problem Solving: 17 1/4 inches

PAGE 109
Review Exercises:
1. 1 1/12
2. 1 1/21
3. 1 5/16
4. 3/10
5. 1/2
6. 1/2
S1. 9 1/4
S2. 8 1/6
1. 11 1/4
2. 8 9/10
3. 10 1/6
4. 12 13/14
5. 5 1/8
6. 16 1/6
7. 7 3/20
8. 14 1/10
9. 5 7/9
10. 8 1/32
Problem Solving: 9 1/4 dollars

PAGE 110
Review Exercises:
1. 3/4
2. 4 9/16
3. 3 1/2
4. 3 2/3
5. 2 1/3
6. 1/2
S1. 3 1/6
S2. 3 5/6
1. 5 1/4
2. 6 7/10
3. 3 8/15
4. 1 7/10
5. 3 11/14
6. 5 3/10
7. 2 4/9
8. 3 9/16
9. 1 11/12
10. 2 11/15
Problem Solving: 16 1/4 dollars

PAGE 111
Review Exercises:
1. 5 3/5
2. 13 1/2
3. 9 3/4
4. 11 14/15
5. 11 1/16
6. 1 1/30
S1. 3 3/4
S2. 2 4/5
1. 4 1/6
2. 4 5/6
3. 2 8/9
4. 2 5/6
5. 5 7/16
6. 3 11/15
7. 8 3/10
8. 3 1/8
9. 4 9/10
10. 6 11/24
Problem Solving: 22 1/4 dollars

PAGE 112
Review Exercises:
1. 5/6
2. 5/8
3. 2 3/4
4. 4
5. 6 5/6
6. 11 1/8
S1. 4/5
S2. 2 1/2
1. 13/24
2. 13/18
3. 3 4/7
4. 13 7/8
5. 18 3/5
6. 4 7/9
7. 3 7/20
8. 3 1/2
9. 23/30
10. 10 1/2
Problem Solving: $204

PAGE 113
Review Exercises:
1. 3/5, 6/10
2. 15/2
3. 21
4. 6 1/5
5. 4/5
6. 1 4/5
S1. 2 1/2
S2. 5 1/4
1. 7/8
2. 1 1/4
3. 17/24
4. 2 1/2
5. 12
6. 6 7/12
7. 12 3/14
8. 1 7/60
9. 1 1/8
10. 8 3/10
Problem Solving: 33 3/4 minutes

PAGE 114
Review Exercises:
1. 5 5/6
2. 3 5/6
3. 1 3/10
4. 4 1/16
5. 7 5/6
6. 2/3
S1. 15/28
S2. 5/9
1. 5/63
2. 1/5
3. 1 7/8
4. 10/27
5. 1 1/15
6. 1 1/15
7. 2 1/4
8. 32/35
9. 2/5
10. 1 13/14
Problem Solving: 37 1/2 pounds

PAGE 115
Review Exercises:
1. 19/8
2. 1/2
3. 4 1/2
4. 13/15
5. 1 2/5
6. 1 4/5
S1. 3/5
S2. 2 5/8
1. 8/15
2. 1 1/4
3. 1 7/20
4. 1 1/3
5. 5/8
6. 1 7/8
7. 3 1/2
8. 5/9
9. 1 1/4
10. 2/3
Problem Solving: $17

PAGE 116
Review Exercises:
1. 3/8
2. 1 2/3
3. 1 1/10
4. 3 1/6
5. 3 3/4
6. 6 2/3
S1. 7/15
S2. 2/3
1. 9/16
2. 1/12
3. 11/18
4. 3 2/3
5. 3/10
6. 2/3
7. 2/5
8. 9/20
9. 1 3/7
10. 1 1/21
Problem Solving: 83 seats

PAGE 117
Review Exercises:
1. 1 2/3
2. 5/8
3. 1 1/6
4. 25/2
5. 6 1/6
6. 7/16
S1. 2/5
S2. 1
1. 7/20
2. 20/27
3. 11/24
4. 4
5. 2 1/4
6. 1/4
7. 1 1/2
8. 1 1/4
9. 1 1/3
10. 11/15
Problem Solving: $13,000

PAGE 118
Review Exercises:
1. 5/6
2. 13 1/2
3. 29/5
4. 12/17
5. 7/20
6. 2 1/6
S1. 12
S2. 3 3/4
1. 12
2. 10
3. 18
4. 2 1/7
5. 18 1/2
6. 1 1/2
7. 2 1/3
8. 7 1/2
9. 35
10. 4 1/5
Problem Solving: 14 girls

PAGE 119
Review Exercises:
1. 1 3/5
2. 8 1/3
3. 10 5/12
4. 2/3
5. 1/4
6. 3/8
S1. 5 3/5
S2. 25
1. 20
2. 9 3/5
3. 6 2/3
4. 45
5. 3 3/4
6. 20
7. 35
8. 16 2/3
9. 4 4/5
10. 3 1/3
Problem Solving: 180 girls

PAGE 120
Review Exercises:
1. 10/21
2. 10/13
3. 6
4. 10
5. 4 2/3
6. 24
S1. 4/9
S2. 5 1/4
1. 1 1/6
2. 7 1/3
3. 11
4. 4 2/5
5. 8 1/4
6. 8 1/8
7. 8 3/4
8. 20
9. 16 1/3
10. 2 5/7
Problem Solving: 36 miles

PAGE 121
Review Exercises:
1. 7/18
2. 3 3/4
3. 16
4. 6 1/6
5. 2 14/15
6. 8 1/10
S1. 4
S2. 6
1. 2 1/4
2. 6 7/8
3. 8 3/4
4. 24
5. 14 2/3
6. 3 3/8
7. 11 1/4
8. 8 2/3
9. 8 1/6
10. 10
Problem Solving: 275 cars

PAGE 122
Review Exercises:
1. 7/12
2. 12
3. 1 2/3
4. 17 1/2
5. 3
6. 3 2/3
S1. 7/18
S2. 4
1. 3/10
2. 1 7/11
3. 27
4. 1 7/8
5. 2 1/10
6. 8/9
7. 11
8. 2 2/15
9. 20
10. 12 3/5
Problem Solving: $60

PAGE 123
Review Exercises:
1. 1/4
2. 13/24
3. 4 1/15
4. 4 2/5
5. 8 7/12
6. 1 7/10
S1. 2 1/2
S2. 7 7/8
1. 1/4
2. 2 1/4
3. 9
4. 9/10
5. 6 3/4
6. 150
7. 5 1/3
8. 22 1/2
9. 5 1/4
10. 21
Problem Solving: 24 students

PAGE 124
Review Exercises:
1. 9/10
2. 5/18
3. 9 1/8
4. 12 1/2
5. 1 3/5
6. 4 1/2
S1. 1 1/2
S2. 2/7
1. 1/7
2. 1 3/5
3. 4/17
4. 1/15
5. 3 1/2
6. 9
7. 1/12
8. 4 1/2
9. 2/11
10. 1/17
Problem Solving: $75

PAGE 125
Review Exercises:
1. 15/2
2. 9/20
3. 2 2/7
4. 49
5. answers vary
6. 1/3, 3/9
S1. 1 1/7
S2. 4/19
1. 1 4/11
2. 1/6
3. 9/28
4. 1/16
5. 7 1/2
6. 6
7. 1/50
8. 4
9. 3/23
10. 1 5/11
Problem Solving: 2 1/4 degrees

PAGE 126
Review Exercises:
1. 1 5/12
2. 1 1/2
3. 4 2/3
4. 5 1/2
5. 1 5/12
6. 1 1/4
S1. 1 3/5
S2. 2 1/4
1. 3 3/4
2. 3/4
3. 10 1/2
4. 11
5. 4 2/3
6. 3 6/11
7. 4/9
8. 3 3/4
9. 3
10. 1 13/15
Problem Solving: 9 pieces

PAGE 127
Review Exercises:
1. 7/12
2. 3
3. 5
4. 1 1/4
5. 1
6. 1 7/9
S1. 1 1/6
S2. 2
1. 2 1/2
2. 6/7
3. 3 1/3
4. 26
5. 3 2/3
6. 3 3/8
7. 8
8. 1 7/8
9. 1 11/16
10. 5
Problem Solving: 27 inches

PAGE 128
Review Exercises:
1. 1 1/8
2. 3/16
3. 1 1/8
4. 3 1/2
5. 4 2/3
6. 2
S1. 6/7
S2. 5/6
1. 11/12
2. 3/5
3. 5/6
4. 9/11
5. 13/15
6. 3/8
7. 23/30
8. 2/11
9. 9/10
10. 11/12
Problem Solving: 19 miles

PAGE 129
Review Exercises:
1. 1 1/2
2. 10
3. 3
4. 9/10
5. 6 7/8
6. 10 1/2
S1. 7/9
S2. 11/20
1. 11/14
2. 10/12
3. 3/4
4. 1/2
5. 2/6
6. 11/20
7. 5/6
8. 7/8
9. 3/5
10. 3/8
Problem Solving: 23

PAGE 130
1. 4/5
2. 4/5
3. 5/6
4. 2 1/5
5. 2 1/3
6. 3 12/25
7. 28/5
8. 64/5
9. 22/3
10. 40
11. 42
12. 60
13. 12
14. 60
15. 12
16. 9/15, 3/5
17. 6/8, 3/4
18. answers vary
19. 2/3
20. 5/6

PAGE 131
1. 1
2. 1 1/2
3. 1 3/20
4. 6 3/8
5. 9 1/2
6. 3/4
7. 2 5/7
8. 4 3/7
9. 4 1/4
10. 4 8/15
11. 3/14
12. 7/32
13. 24
14. 3 1/2
15. 8 3/4
16. 2 1/2
17. 14
18. 1 1/2
19. 2 8/11
20. 2 3/11

PAGE 132
1. 6/7
2. 3/5
3. 4/5
4. 3 1/4
5. 1 7/12
6. 1 3/7
7. 17/8
8. 26/3
9. 17/16
10. 9
11. 16
12. 49
13. 30
14. 24
15. 30
16. 2/6, 1/3
17. 2/8, 1/4
18. answers vary
19. 4/7
20. 4/5

PAGE 133
1. 4/5
2. 1 1/2
3. 1 3/10
4. 6 1/15
5. 8 2/21
6. 4/5
7. 3 3/4
8. 3 5/8
9. 4 11/20
10. 4 2/3
11. 4/9
12. 1/6
13. 21
14. 2 3/5
15. 6
16. 1 1/2
17. 7 1/3
18. 1 13/14
19. 2 6/17
20. 1 1/2

Decimals and Percents—Solutions

PAGE 136
Review Exercises
1. 97
2. 215
3. 451
4. 620
5. 109
6. 624
S1. three and seven tenths
S2. twelve and ninteen thousandths
1. eighty-seven hundredths
2. five and six thousandths
3. one hundred fifteen and seven tenths
4. seventy-eight and seven hundredths
5. six and three thousand nine hundred twelve ten-thousandths
6. eighty-five thousandths
7. seven and thirty-six hundredths
8. nine and two thousandths
9. sixty-one hundredths
10. two and three hundred thirty-three thousandths
Problem Solving: $180

PAGE 137
Review Exercises:
1. 1,408
2. 749
3. 523
4. 115
5. 224
6. 1,260
S1. three and six thousandths
S2. one hundred seventy-six ten-thousandths
1. eight tenths
2. three and five ten-thousandths
3. seventy-six and eight tenths
4. seven and eight thousandths
5. five and one hundred thirty-eight thousandths
6. fifteen thousandths
7. five and eighty-two hundredths
8. four and three hundredths
9. eighty-six hundredths
10. four and two hundred twenty-four thousandths
Problem Solving: 125 students

PAGE 138
Review Exercises:
1. 779
2. two and seven thousandths
3. 6,382
4. forty-two and sixteen thousandths
5. 327
6. nineteen thousandths
S1. 5.03
S2. 436.11
1. 7.4
2. 22.015
3. .0352
4. 74.043
5. .00005
6. .000016
7. 9.045
8. 20.0033
9. 86.9
10. 86.000009
Problem Solving: 360 crayons

PAGE 139
Review Exercises:
1. two and nine hundredths
2. two and nine thousandths
3. 2.04
4. 682
5. .015
6. 13,182
S1. 7.062
S2. .02009
1. 8.09
2. 12.00041
3. .0049
4. 97.000513
5. .048
6. 52.8
7. 5.496
8. 3.005
9. 12.00033
10. 116.05
Problem Solving: 24 students

PAGE 140
Review Exercises:
1. two and seven hundredths
2. seven and seventeen thousandths
3. 7.006
4. .000032
5. seventeen ten-thousandths
6. 5.0011
S1. 5.6
S2. 8.009
1. 21.16
2. .16
3. 14.017
4. 119.00016
5. .0021
6. .00196
7. 4.032
8. 3.324
9. 4.000017
10. .0019
Problem Solving: 450 mph

PAGE 141
Review Exercises:
1. 10.014
2. two and nine hundredths
3. .065
4. 10.00015
5. 107
6. 564
S1. 9.017
S2. 9.00017
1. 42.196
2. .072
3. 48.008
4. 16.00195
5. .016
6. 16.000119
7. 4.0038
8. 3.0176
9. .071
10. 6.0053
Problem Solving: 384 miles

Decimals and Percents—Solutions

PAGE 142
Review Exercises:
1. 9.07
2. 1.00135
3. .00017
4. twenty-one thousandths
5. three and nineteen hundreths
6. 6.0013

S1. $3 \frac{5}{100}$
S2. $16 \frac{17}{1,000}$
1. $45 \frac{19}{10,000}$
2. $\frac{5}{100,000}$
3. $7 \frac{16}{1,000,000}$
4. $7 \frac{196}{1,000}$
5. $79 \frac{6}{10}$
6. $\frac{7,632}{100,000}$
7. $14 \frac{7}{1,000,000}$
8. $16 \frac{24}{1,000}$
9. $17 \frac{145}{1,000,000}$
10. $\frac{96}{100,000}$

Problem Solving: $274

PAGE 143
Review Exercises:
1. 11.06
2. $\frac{6}{1,000}$
3. .019
4. $7 \frac{6}{1,000}$
5. $6 \frac{3}{10}$
6. .072

S1. $\frac{16}{10,000}$
S2. $9 \frac{125}{100,000}$
1. $7 \frac{9}{100,000}$
2. $\frac{16}{1,000}$
3. $7 \frac{29}{100}$
4. $6 \frac{2}{100,000}$
5. $87 \frac{3}{10}$
6. $5 \frac{72}{10,000}$
7. $15 \frac{6}{1,000,000}$
8. $42 \frac{1}{10}$
9. $163 \frac{137}{10,000}$
10. $\frac{11,234}{100,000}$

Problem Solving: 6 boxes, 4 left

PAGE 144
Review Exercises:
1. $7 \frac{16}{10,000}$
2. 7.015
3. 5.6
4. six and thirteen thousandths
5. three and seven thousandths
6. $72 \frac{12}{10,000}$

S1. <
S2. >
1. <
2. >
3. <
4. >
5. >
6. <
7. <
8. >
9. <
10. <

Problem Solving: 8 gallons, $24

PAGE 145
Review Exercises:
1. $\frac{129}{10,000}$
2. seven and ninety-two thousandths
3. one hundred thirty-five ten thousandths
4. 17.006
5. 15.0071
6. $3 \frac{96}{1,000}$

S1. <
S2. <
1. <
2. >
3. >
4. >
5. <
6. <
7. >
8. >
9. >
10. <

Problem Solving: $181

PAGE 146
Review Exercises:
1. $52 \frac{623}{1,000}$
2. 75.009
3. 4,368
4. 50.019
5. six and twenty-three hundredths
6. 6,784

S1. 18.82
S2. 8.84
1. 41.353
2. 14.431
3. 29.673
4. 1.7
5. 19.26
6. 146.983
7. 1.7
8. 25.044
9. 42.055
10. 31.09

Problem Solving: 18.62 inches

PAGE 147
Review Exercises:
1. seven and nineteen hundredths
2. 14.503
3. 7.0005
4. $7 \frac{632}{10,000}$
5. nineteen thousandths
6. 1.9

S1. 25.73
S2. 13.58
1. 50.103
2. 18.542
3. 34.976
4. 1.75
5. 107.36
6. 103.306
7. 2.3
8. 86.415
9. 56.602
10. 25.55

Problem Solving: $15

Decimals and Percents—Solutions

PAGE 148
Review Exercises:
1. 19.557
2. 14.06
3. 7.0125
4. $6\ ^{12}/_{1,000}$
5. two and seven hundredths
6. seven and five thousandths

S1. 13.84
S2. 7.48
1. 5.894
2. 2.293
3. 1.16
4. 11.13
5. .053
6. 3.9577
7. 2.373
8. 3.684
9. 8.628
10. 26.765
Problem Solving: 2.7°

PAGE 149
Review Exercises:
1. 2.6
2. 12.17
3. seventy-two hundredths
4. seventy-two ten-thousandths
5. $75\ ^{6}/_{10,000}$
6. 97.03

S1. 55.62
S2. 73.04
1. 4.003
2. 7.187
3. 3.478
4. 36.34
5. .193
6. 4.8884
7. 4.036
8. 87.976
9. 2.286
10. 57.335
Problem Solving: $279

PAGE 150
Review Exercises:
1. 17.79
2. 5.058
3. 17.36
4. 12.22
5. .075
6. two and fifty-eight thousandths

S1. 18.38
S2. 3.687
1. 24.164
2. 6.17
3. 16.57
4. 8.24
5. 2.08
6. 19.56
7. 2.2
8. 4.487
9. 10.396
10. 25.7
Problem Solving: 5.5 pounds

PAGE 151
Review Exercises:
1. .000007
2. seven millionths
3. 79.22
4. 6.0022
5. 4.64
6. 2.335

S1. 49.417
S2. 6.764
1. 17.19
2. 2.2
3. 26.73
4. 3.234
5. 93.106
6. 2.063
7. 2.1
8. 66.218
9. 9.33
10. 73.001
Problem Solving: 347.95 miles

PAGE 152
Review Exercises:
1. 170
2. 5,551
3. 11,052
4. 20.69
5. 4.563
6. 16.5

S1. 7.38
S2. 36.8
1. 1.929
2. 14.64
3. 6.88
4. 56.64
5. 22.4
6. 55.2
7. 328.09
8. 10.024
9. 29.93
10. 1.242
Problem Solving: 38 tons

PAGE 153
Review Exercises:
1. 12.78
2. 32.2
3. 36.065
4. .736
5. 4.733
6. 19.747

S1. 17.35
S2. 101.2
1. 2.892
2. 22.32
3. 12.42
4. 86.88
5. 67.2
6. 79.35
7. 56.992
8. 17.82
9. 570
10. 3.288
Problem Solving: $100.75

PAGE 154
Review Exercises:
1. 15.587
2. 5.656
3. 5.75
4. 3.84
5. 5.63
6. 22.636
S1. 2.52
S2. 7.776
1. 11.52
2. .4598
3. .21186
4. .874
5. .1421
6. 9.7904
7. .0024
8. 19.04
9. 149.225
10. .0738
Problem Solving: $8.00

PAGE 155
Review Exercises:
1. 38.52
2. three and twenty-six ten-thousandths
3. 3 $7/1{,}000$
4. 2.011
5. sixteen hundred-thousandths
6. 1.058
S1. .2268
S2. 13.692
1. 24.38
2. .6897
3. .18744
4. .2442
5. 1.242
6. 1.18209
7. .0042
8. 2.752
9. 15.2334
10. .1016
Problem Solving: $87.50

PAGE 156
Review Exercises:
1. 6.42
2. 3.621
3. .000006
4. 16
5. 1.857
6. $17.70
S1. 32
S2. 7,390
1. 93.6
2. 72,600
3. 160
4. 736.2
5. 7,280
6. 70
7. 37.6
8. 390
9. 73.3
10. 76.3
Problem Solving: $18,500

PAGE 157
Review Exercises:
1. two and seven hundred sixty-three thousandths
2. .00007
3. 4.45
4. .00016
5. .3615
6. 53.8
S1. 370
S2. 5,360
1. 327
2. 56,700
3. 190
4. 736.4
5. 976
6. 750
7. 73
8. 387
9. 8,330
10. 84.2
Problem Solving: 2,500 pounds

PAGE 158
Review Exercises:
1. 139.2
2. 9
3. 3,250
4. 4.236
5. .000009
6. 21.573
S1. 2.394
S2. 15.228
1. 3.22
2. .464
3. .48508
4. 26
5. 3.156
6. 2,630
7. .0108
8. 3.575
9. 5.511
10. .04221
Problem Solving: $72

PAGE 159
Review Exercises:
1. 11.75
2. 4.766
3. 75.382
4. 12.24
5. 14.88
6. 2,365
S1. 25.56
S2. .3596
1. 96.6
2. 2.961
3. .00627
4. 360
5. 89.28
6. .000005
7. 138.072
8. 42,900
9. 34.72
10. .02898
Problem Solving: $13.52

Decimals and Percents—Solutions

PAGE 160
Review Exercises:
1. 22 r2
2. 303
3. 111
4. 361 r21
5. 132
6. 203
S1. .44
S2. 1.8
1. 19.7
2. 3.21
3. .57
4. 3.7
5. 6.08
6. 40.9
7. .324
8. .16
9. 2.12
10. 2.04
Problem Solving: $1.95

PAGE 161
Review Exercises:
1. 4.4
2. .345
3. 1.63
4. 5.483
5. .864
6. 78.29
S1. .232
S2. 2.68
1. .51
2. 4.06
3. .56
4. 5.6
5. 2.304
6. .23
7. .63
8. .16
9. .123
10. 4.08
Problem Solving: 12.2 seconds

PAGE 162
Review Exercises:
1. 7.38
2. .301
3. 270
4. 38.8
5. 279.66
6. $2\ ^{1,762}/_{10,000}$
S1. .0027
S2. .019
1. .0007
2. .012
3. .056
4. .036
5. .043
6. .063
7. .023
8. .022
9. .003
10. .068
Problem Solving: $33

PAGE 163
Review Exercises:
1. two and three thousandths
2. 110.213
3. 66.987
4. 436.1
5. 13.3
6. .172
S1. .055
S2. .038
1. .0004
2. .002
3. .072
4. .046
5. .034
6. .054
7. .057
8. .044
9. .0008
10. .034
Problem Solving: $5.25

PAGE 164
Review Exercises:
1. 67.8
2. .02052
3. 360
4. .003
5. .011
6. 6.00015
S1. .34
S2. .06
1. .065
2. .62
3. 2.05
4. .15
5. .04
6. .04
7. .12
8. 1.575
9. .006
10. .418
Problem Solving: $10.13

PAGE 165
Review Exercises:
1. $1.59
2. $63.20
3. .065
4. 47.48
5. 18.518
6. .08
S1. .074
S2. 2.25
1. .085
2. .078
3. .06
4. .16
5. .638
6. .15
7. 1.45
8. .185
9. 1.35
10. .04
Problem Solving: $1.81

PAGE 166
Review Exercises:
1. .095
2. .34
3. .0002
4. 5.0013
5. $^{135}/_{100,000}$
6. 2.2
S1. 3.9
S2. .24
1. 8
2. 170
3. .42
4. 80
5. 5.4
6. 3.2
7. 3.7
8. 3.2
9. 57.5
10. 9.2
Problem Solving: $2.13

PAGE 167
Review Exercises:
1. 2.07
2. 5.91
3. 4.104
4. .65
5. .034
6. .43
S1. 3.125
S2. 508
1. 1.25
2. .03
3. 5.08
4. 230
5. 20
6. 6.5
7. .33
8. 50
9. 9.2
10. 2.4
Problem Solving: 55 tiles

PAGE 168
Review Exercises:
1. 1.6
2. 15
3. .34
4. 300
5. 10.1
6. 5.2
S1. .75
S2. .625
1. .6
2. .25
3. .4
4. .875
5. .55
6. .52
7. .625
8. .2
9. .2
10. .7
Problem Solving: $420

PAGE 169
Review Exercises:
1. 2.3
2. 29.09
3. .2898
4. 4.877
5. 4
6. 50
S1. .375
S2. .75
1. .8
2. .1875
3. .44
4. .15
5. .75
6. .2
7. .26
8. .7
9. .28
10. .4375
Problem Solving: $2.76

PAGE 170
Review Exercises:
1. 10.48
2. 9.84
3. 123.6
4. 22.12
5. .0132
6. .38
S1. .005
S2. 40
1. .76
2. .074
3. .025
4. 4.5
5. .24
6. 284
7. 33.1
8. 16.2
9. .4
10. .625
Problem Solving: $9.50

PAGE 171
Review Exercises:
1. 19.7
2. .16
3. .003
4. 1.575
5. 54
6. 3.2
S1. .625
S2. 70
1. .0078
2. .78
3. 7.3
4. 5
5. 21.25
6. 286
7. .23
8. .65
9. .75
10. .25
Problem Solving: $12,000

Decimals and Percents—Solutions

PAGE 172

1. 16.336
2. 13.87
3. 34.5
4. 27.9
5. 5.548
6. 70.51
7. 19.04
8. 128.8
9. 1.274
10. 1.74038
11. 3,190
12. 320
13. 1.98
14. .158
15. .32
16. 300
17. .03
18. 2.4
19. .625
20. .5625

PAGE 173

1. 41.716
2. 16.22
3. 2.9
4. 71.28
5. 52.877
6. .084
7. 4.685
8. 84
9. 3.528
10. .14283
11. 410
12. 3,762
13. .79
14. .0064
15. .25
16. 6,000
17. 470
18. 2.12
19. .125
20. .875

PAGE 174
Review Exercises:
1. .17
2. .9
3. 7/100
4. 7/10
5. six and seven hundredths
6. ten and nine tenths
S1. 17%
S2. 90%
1. 6%
2. 99%
3. 30%
4. 64%
5. 67%
6. 1%
7. 70%
8. 14%
9. 80%
10. 62%
Problem Solving: 12 gallons, $39

PAGE 175
Review Exercises:
1. .017
2. 500
3. 340
4. 4.284
5. 69.27
6. 4.136
S1. 70%
S2. 3%
1. 19%
2. 87%
3. 60%
4. 63%
5. 19%
6. 2%
7. 48%
8. 14%
9. 50%
10. 98%
Problem Solving: $172

PAGE 176
Review Exercises:
1. 70%
2. 72%
3. 400
4. .625
5. .6
6. .000006
S1. 37%
S2. 70%
1. 93%
2. 2%
3. 20%
4. 9%
5. 60%
6. 66.5%
7. 89%
8. 60%
9. 33.4%
10. 80%
Problem Solving: 9.6 fluid ounces

PAGE 177
Review Exercises:
1. 3 $196/10,000$
2. twenty-one ten-thousandths
3. 7.0007
4. 3.03
5. 4
6. .22
S1. 9%
S2. 34.8%
1. 90%
2. 9%
3. 70%
4. 9.7%
5. 60%
6. .7%
7. 87%
8. 30%
9. 44.5%
10. 40%
Problem Solving: 120 students

Decimals and Percents—Solutions

PAGE 178
Review Exercises:
1. 7.11
2. 2.22
3. 19.24
4. 3.19
5. 2.463
6. 60.387
S1. .2, 1/5
S2. .09, 9/100
1. .16, 4/25
2. .06, 3/50
3. .75, 3/4
4. .4, 2/5
5. .01, 1/100
6. .45, 9/20
7. .12, 3/25
8. .05, 1/20
9. .5, 1/2
10. .13, 13/100
Problem Solving: 1/4

PAGE 179
Review Exercises:
1. 63.5
2. 18.375
3. .00024
4. .075
5. .202
6. 2.6
S1. .5, 1/2
S2. .05, 1/20
1. .08, 2/25
2. .8, 4/5
3. .24, 6/25
4. .11, 11/100
5. .02, 1/50
6. .7, 7/10
7. .09, 9/100
8. .9, 9/10
9. .17, 17/100
10. .14, 7/50
Problem Solving: $880

PAGE 180
Review Exercises:
1. .8
2. .07
3. 1/4
4. 109.2
5. 128
6. 18
S1. 17.5
S2. 150
1. 4.32
2. 51
3. 15
4. 112.5
5. 32
6. 80
7. 10
8. 216
9. 112.5
10. 13.2
Problem Solving:
    160 miles per hour

PAGE 181
Review Exercises:
1. .4
2. .4
3. .65
4. 40.5
5. .375
6. three and sixteen
   ten-thousandths
S1. 24
S2. 240
1. 3.2
2. 385
3. 38.5
4. 50
5. 50
6. 186
7. 11
8. 240
9. 120
10. 6
Problem Solving: 84

PAGE 182
Review Exercises:
1. 7.2
2. 288
3. 10
4. .75
5. 1/2
6. 22.5
S1. 6 correct
S2. 30 absent
1. $84 in bank
2. $2,700 saved
3. 18 incorrect
4. $600 on food
5. 12 girls
6. $40,000 down payment
7. $5,600 value
8. $3.50, $53.50 total
9. 150 students
10. $1,500 on food & housing
Problem Solving: $297

PAGE 183
Review Exercises:
1. 48
2. 4.8
3. 180
4. .2
5. 2.3
6. 1.45
S1. 32 passed
S2. 18 girls
1. 540 cookies
2. 72 white marbles
3. 480 students
4. $3.60 saved
5. $3,600 down payment
6. $480 spent
7. $9.00 tip
8. $480 for rent & car
9. 30 did not pass
10. 360 boxes, 240 girls
Problem Solving: $7.50

Decimals and Percents—Solutions

PAGE 184
Review Exercises:
1. 4.9
2. 7.2
3. 7.2
4. 8.44
5. 7.009
6. .09
S1. 20%
S2. 80%
1. 60%
2. 50%
3. 10%
4. 75%
5. 75%
6. 60%
7. 25%
8. 80%
9. 75%
10. 20%
Problem Solving: 144 people

PAGE 185
Review Exercises:
1. 93.29
2. 14.85
3. 1.519
4. .8
5. 11.1
6. .21
S1. 80%
S2. 75%
1. 50%
2. 70%
3. 25%
4. 20%
5. 25%
6. 25%
7. 62.5%
8. 50%
9. 25%
10. 37.5%
Problem Solving: 75%

PAGE 186
Review Exercises:
1. 46.5
2. 24
3. .75
4. 50%
5. 135
6. 600
S1. 25%
S2. 75%
1. 25%
2. 80%
3. 50%
4. 90%
5. 60%
6. 75%
7. 75%
8. 75%
9. 80%
10. 95%
Problem Solving: 320 cows

PAGE 187
Review Exercises:
1. 7.28
2. 3/1,000
3. 24
4. .9
5. 100
6. 3.328
S1. 75%
S2. 20%
1. 40%
2. 75%
3. 90%
4. 45%
5. 25%
6. 75%
7. 75%
8. 60%
9. 90%
10. 52%
Problem Solving: 36 correct

PAGE 188
Review Exercises:
1. 112
2. 3
3. 60%
4. 60%
5. 47.1%
6. 72.00601
S1. 60% correct
S2. 60% girls
1. 80% correct
2. 25% into savings
3. 25% lost
4. 60% caught
5. 90 percent
6. 20% absent
7. 48% won
8. 40% are sixth graders
9. 98% correct
10. 75% strikes
Problem Solving: $453.60

PAGE 189
Review Exercises:
1. 55%
2. 70%
3. 8
4. 20
5. 500
6. 500
S1. 75% did not get A's
S2. missed 25%
1. 80% won
2. 20% lost
3. 60% boys
4. 60% into savings
5. 60% take bus
6. 40% are goldfish
7. 25% missed
8. 90% have a computer
9. 75% completed
10. 30% have a pet
Problem Solving: $50.15

Decimals and Percents—Solutions

PAGE 190
Review Exercises:
1. 3.2
2. 32
3. 75%
4. 90%
5. 50.04
6. 200
S1. 20
S2. 30
1. 48
2. 80
3. 25
4. 4
5. 15
6. 20
7. 60
8. 75
9. 45
10. 125
Problem Solving: 90%

PAGE 191
Review Exercises:
1. 7.2
2. 24
3. 25%
4. 20%
5. 16
6. 25
S1. 200
S2. 40
1. 20
2. 50
3. 60
4. 50
5. 25
6. 250
7. 35
8. 36
9. 80
10. 150
Problem Solving: 225 cakes

PAGE 192
Review Exercises:
1. 12
2. 25%
3. 15
4. .8
5. $7 \frac{9}{1,000}$
6. 6.017
S1. 15 games
S2. $600 earned
1. 120 stamps
2. 12 shots taken
3. 500 students in school
4. 80
5. $160 earned
6. 125 cows in herd
7. 140
8. 20 problems on test
9. 80 tried out
10. 30 problems on test
Problem Solving: $16.28

PAGE 193
Review Exercises:
1. 92.943
2. 1.414
3. .3696
4. 4,500
5. .09
6. 3,200
S1. 50 questions
S2. 125 took test
1. 15 in class
2. 12 pitches thrown
3. 125
4. 20 marbles in bag
5. 75 in class
6. 15
7. $40 bill
8. $24,000 earnings
9. $200 for bike
10. 5
Problem Solving: $16.74

PAGE 194
Review Exercises:
1. .008
2. .9
3. .625
4. .1065
5. 15.73
6. 1.5
S1. 20%
S2. 20
1. 42
2. 21
3. 25%
4. 25
5. 20
6. 60
7. 75%
8. 16
9. 60
10. 20%
Problem Solving: 10 problems

PAGE 195
Review Exercises:
1. 7.2
2. 1.715
3. 10
4. .26
5. .25
6. .26
S1. 288
S2. 35
1. 5%
2. 25.6
3. 80
4. 24
5. 24
6. 75%
7. 6
8. 320
9. 20%
10. 30
Problem Solving: 75%

Decimals and Percents—Solutions

PAGE 196

Review Exercises:

1. .00072
2. 2 $^{19}/10{,}000$
3. 60%
4. 15
5. .021
6. 8

S1. 20 correct

S2. 75% made shots

1. earnings $25
2. $1,600 down payment
3. 30
4. 10% lost
5. 250 in herd
6. 25
7. $240 saved
8. 180 boys
9. 20%
10. $210 into bank

Problem Solving: $57,600

PAGE 197

Review Exercises:

1. 32.744
2. 2.358
3. 21.98
4. .01248
5. .48
6. 8.1

S1. 30

S2. 30

1. 20%
2. 90%
3. 30 absent
4. 18 caught
5. 20% are red
6. 75%
7. 250 in school
8. $25 regular price
9. 60% of test
10. $691.20 total

Problem Solving: 97

PAGE 198

1. 19%
2. 7%
3. 90%
4. 27%
5. 30%
6. .03, 3/100
7. .16, 4/25
8. .9, 9/10
9. 16.8
10. 315
11. 96
12. 20%
13. 40%
14. 25
15. 120
16. 75%
17. 120 girls
18. 80% correct
19. 24 on test
20. lost 13 games

PAGE 199

1. 70%
2. 7%
3. 10%
4. 5%
5. 50%
6. .7, 7/10
7. .02, 1/50
8. .15, 3/20
9. 220
10. 100
11. 5.4
12. 40
13. 30%
14. 25
15. 16.8
16. 96
17. $1,200 withdrawn
18. 40 on team
19. $1,920 sales tax
20. 80% spent

A

absolute value   The distance of a number from 0 on the number line. The absolute value is always positive.

acute angle   An angle with a measure of less than 90 degrees.

adjacent   Next to.

algebraic expression   A mathematical expression that contains at least one variable.

angle   Any two rays that share an endpoint will form an angle.

associative properties   For any a, b, c:
addition: $(a + b) + c = a + (b + c)$
multiplication: $(ab)c = a(bc)$

B

base   The number being multiplied. In an expression such as 4^2, 4 is the base.

C

coefficient   A number that multiplies the variable. In the term 7x, 7 is the coefficient of x.

commutative properties   For any a, b:
addition: $a + b = b + a$
multiplication: $ab = ba$

complementary angles   Two angles that have measures whose sum is 90 degrees.

congruent   Two figures having exactly the same size and shape.

coordinate plane   The plane which contains the x- and y-axes. It is divided into 4 quadrants. Also called coordinate system and coordinate grid.

coordinates   An ordered pair of numbers that identify a point on a coordinate plane.

D

data   Information that is organized for analysis.

degree   A unit that is used in measuring angles.

denominator   The bottom number of a fraction that tells the number of equal parts into which a whole is divided.

disjoint sets   Sets that have no members in common. {1,2,3} and {4,5,6} are disjoint sets.

Glossary

distributive property   For real numbers a, b, and c: a(b + c) = ab + ac.

E

element of a set   Member of a set.

empty set   The set that has no members. Also called the null set and written Ø or { }.

equation   A mathematical sentence that contains an equal sign (=) and states that one expression is equal to another expression.

equivalent   Having the same value.

exponent   A number that indicates the number of times a given base is used as a factor. In the expression n^2, 2 is the exponent.

expression   Variables, numbers, and symbols that show a mathematical relationship.

extremes of a proportion   In the proportion $\frac{a}{b} = \frac{c}{d}$, a and d are the extremes.

F

factor   An integer that divides evenly into another.

finite   Something that is countable.

formula   A general mathematical statement or rule. Used often in algebra and geometry.

function   A set of ordered pairs that pairs each x-value with one and only one y-value. (0,2), (-1,6), (4,-2), (-3,4) is a function.

G

graph   To show points named by numbers or ordered pairs on a number line or coordinate plane. Also, a drawing to show the relationship between sets of data.

greatest common factor   The largest common factor of two or more numbers. Also written GCF. The greatest common factor of 15 and 25 is 5.

grouping symbols   Symbols that indicate the order in which mathematical operations should take place. Examples include parentheses ( ), brackets [ ], braces { }, and fraction bars —— .

H

hypotenuse   The side opposite the right angle in a right triangle.

I

identity properties of addition and multiplication   For any real number a:
addition: $a + 0 = 0 + a = a$
multiplication: $1 \times a = a \times 1 = a$

inequality   A mathematical sentence that states one expression is greater than or less than another. Inequality symbols are read as follows: $<$ less than
$\leq$ less than or equal to
$>$ greater than
$\geq$ greater than or equal to

infinite   Having no boundaries or limits. Uncountable.

integers   Numbers in a set. ...-3, -2, -1, 0, 1, 2, 3...

intersection of sets   If A and B are sets, then A intersection B is the set whose members are included in both sets A and B, and is written $A \cap B$. If set A = {1,2,3,4} and set B = {1,3,5}, then $A \cap B$ = {1,3}

inverse properties of addition and multiplication   For any number a:
addition: $a + -a = 0$
multiplication: $a \times 1/a = 1$ $(a \neq 0)$

inverse operations   Operations that "undo" each other. Addition and subtraction are inverse operations, and multiplication and division are inverse operations.

L

least common multiple   The least common multiple of two or more whole numbers is the smallest whole number, other than zero, that they all divide into evenly. Also written as LCM. The least common multiple of 12 and 8 is 24.

linear equation   An equation whose graph is a straight line.

M

mean   In statistics, the sum of a set of numbers divided by the number of elements in the set. Sometimes referred to as average.

means of a proportion   In the proportion $\frac{a}{b} = \frac{c}{d}$, b and c are the means.

median   In statistics, the middle number of a set of numbers when the numbers are arranged in order of least to greatest. If there are two middle numbers, find their mean.

mode   In statistics, the number that appears most frequently. Sometimes there is no mode. There may also be more than one mode.

multiple   The product of a whole number and another whole number.

Glossary

N

natural numbers   Numbers in the set 1, 2, 3, 4,... Also called counting numbers.

negative numbers   Numbers that are less than zero.

null set   The set that has no members. Also called the empty set and written Ø or { }.

number line   A line that represents numbers as points.

numerator   The top part of a fraction.

O

obtuse angle   An angle whose measure is greater than 90° and less than 180°.

opposites   Numbers that are the same distance from zero, but are on opposite sides of zero on a number line. 4 and -4 are opposites.

order of operations   The order of steps to be used when simplifying expressions.
1. Evaluate within grouping symbols.
2. Eliminate all exponents.
3. Multiply and divide in order from left to right.
4. Add and subtract in order from left to right.

ordered pair   A pair of numbers (x,y) that represent a point on the coordinate plane. The first number is the x-coordinate and the second number is the y-coordinate.

origin   The point where the x-axis and the y-axis intersect in a coordinate plane. Written as (0,0).

outcome   One of the possible events in a probability situation.

P

parallel lines   Lines in a plane that do not intersect. They stay the same distance apart.

percent   Hundredths or per hundred. Written %.

perimeter   The distance around a figure.

perpendicular lines   Lines in the same plane that intersect at a right (90°) angle.

pi   The ratio of the circumference of a circle to its diameter. Written π. The approximate value for π is 3.14 as a decimal and $\frac{22}{7}$ as a fraction.

plane   A flat surface that extends infinitely in all directions.

point   An exact position in space. Points also represent numbers on a number line or coordinate plane.

positive number   Any number that is greater than 0.

power   An exponent.

prime number   A whole number greater than 1 whose only factors are 1 and itself.

probability   What chance, or how likely it is for an event to occur. It is the ratio of the ways a certain outcome can occur and the number of possible outcomes.

proportion   An equation that states that two ratios are equal. $\frac{4}{8} = \frac{2}{4}$ is a proportion.

Pythagorean theorem   In a right triangle, if c is the hypotenuse, and a and b are the other two legs, then $a^2 + b^2 = c^2$.

Q

quadrant   One of the four regions into which the x-axis and y-axis divide a coordinate plane.

R

range   The difference between the greatest number and the least number in a set of numbers.

ratio   A comparison of two numbers using division. Written a:b, a to b, and a/b.

reciprocals   Two numbers whose product is 1. $\frac{2}{3}$ and $\frac{3}{2}$ are reciprocals because $\frac{2}{3} \times \frac{3}{2} = 1$.

reduce   To express a fraction in its lowest terms.

relation   Any set of ordered pairs.

right angle   An angle that has a measure of 90°.

rise   The change in y going from one point to another on a coordinate plane. The vertical change.

run   The change in x going from one point to another on a coordinate plane. The horizontal change.

S

scientific notation   A number written as the product of a numbers between 1 and 10 and a power of ten. In scientific notation, $7{,}000 = 7 \times 10^3$.

set   A well-defined collection of objects.

slope   Refers to the slant of a line. It is the ratio of rise to run.

Glossary

solution   A number that can be substituted for a variable to make an equation true.

square root   Written $\sqrt{}$. The $\sqrt{36} = 6$ because $6 \times 6 = 36$.

statistics   Involves data that is gathered about people or things and is used for analysis.

subset   If all the members of set A are members of set B, then set A is a subset of set B. Written $A \subset B$. If set A = {1,2,3} and set B = {0,1,2,3,5,8}, set A is a subset of set B because all of the members of a set A are also members of set B.

U

union of sets   If A and B are sets, the union of set A and set B is the set whose members are included in set A, or set B, or both set A and set B. A union B is written $A \cup B$. If set = {1,2,3,4} and set B = {1,3,5,7}, then $A \cup B$ = {1,2,3,4,5,7}.

universal set   The set which contains all the other sets which are under consideration.

V

variable   A letter that represents a number.

Venn diagram   A type of diagram that shows how certain sets are related.

vertex   The point at which two lines, line segments, or rays meet to form an angle.

W

whole number   Any number in the set 0, 1, 2, 3, 4...

X

x-axis   The horizontal axis on a coordinate plane.

x-coordinate   The first number in an ordered pair. Also called the abscissa.

Y

y-axis   The vertical axis on a coordinate plane.

y-coordinate   The second number in an ordered pair. Also called the ordinate.

Important Symbols

$<$	less than	π	pi
$\leq$	less than or equal to	$\{\ \}$	set
$>$	greater than	$\|\ \|$	absolute value
$\geq$	greater than or equal to	$.\overline{n}$	repeating decimal symbol
$=$	equal to	$1/a$	the reciprocal of a number
$\neq$	not equal to	$\%$	percent
$\cong$	congruent to	(x,y)	ordered pair
$(\ )$	parenthesis	$\perp$	perpendicular
$[\ ]$	brackets	$\|\ \|$	parallel to
$\{\ \}$	braces	$\angle$	angle
$...$	and so on	$\in$	element of
$\cdot$ or $\times$	multiply	$\notin$	not an element of
∞	infinity	$\cap$	intersection
a^n	the nth power of a number	$\cup$	union
$\sqrt{\ \ }$	square root	$\subset$	subset of
$\varnothing, \{\ \}$	the empty set or null set	$\not\subset$	not a subset of
$\therefore$	therefore	$\triangle$	triangle
$\circ$	degree		

Multiplication Table

x	2	3	4	5	6	7	8	9	10	11	12
2	4	6	8	10	12	14	16	18	20	22	24
3	6	9	12	15	18	21	24	27	30	33	36
4	8	12	16	20	24	28	32	36	40	44	48
5	10	15	20	25	30	35	40	45	50	55	60
6	12	18	24	30	36	42	48	54	60	66	72
7	14	21	28	35	42	49	56	63	70	77	84
8	16	24	32	40	48	56	64	72	80	88	96
9	18	27	36	45	54	63	72	81	90	99	108
10	20	30	40	50	60	70	80	90	100	110	120
11	22	33	44	55	66	77	88	99	110	121	132
12	24	36	48	60	72	84	96	108	120	132	144

Commonly Used Prime Numbers

2	3	5	7	11	13	17	19	23	29
31	37	41	43	47	53	59	61	67	71
73	79	83	89	97	101	103	107	109	113
127	131	137	139	149	151	157	163	167	173
179	181	191	193	197	199	211	223	227	229
233	239	241	251	257	263	269	271	277	281
283	293	307	311	313	317	331	337	347	349
353	359	367	373	379	383	389	397	401	409
419	421	431	433	439	443	449	547	461	463
467	479	487	491	499	503	509	521	523	541
547	557	563	569	571	577	587	593	599	601
607	613	617	619	631	641	643	647	653	659
661	673	677	683	691	701	709	719	727	733
739	743	751	757	761	769	773	787	797	809
811	821	823	827	829	839	853	857	859	863
877	881	883	887	907	911	919	929	937	941
947	953	967	971	977	983	991	997	1009	1013

Squares and Square Roots

No.	Square	Square Root	No.	Square	Square Root	No.	Square	Square Root
1	1	1.000	51	2,601	7.141	101	10201	10.050
2	4	1.414	52	2,704	7.211	102	10,404	10.100
3	9	1.732	53	2,809	7.280	103	10,609	10.149
4	16	2.000	54	2,916	7.348	104	10,816	10.198
5	25	2.236	55	3,025	7.416	105	11,025	10.247
6	36	2.449	56	3,136	7.483	106	11,236	10.296
7	49	2.646	57	3,249	7.550	107	11,449	10.344
8	64	2.828	58	3,364	7.616	108	11,664	10.392
9	81	3.000	59	3,481	7.681	109	11,881	10.440
10	100	3.162	60	3,600	7.746	110	12,100	10.488
11	121	3.317	61	3,721	7.810	111	12,321	10.536
12	144	3.464	62	3,844	7.874	112	12,544	10.583
13	169	3.606	63	3,969	7.937	113	12,769	10.630
14	196	3.742	64	4,096	8.000	114	12,996	10.677
15	225	3.873	65	4,225	8.062	115	13,225	10.724
16	256	4.000	66	4,356	8.124	116	13,456	10.770
17	289	4.123	67	4,489	8.185	117	13,689	10.817
18	324	4.243	68	4,624	8.246	118	13,924	10.863
19	361	4.359	69	4,761	8.307	119	14,161	10.909
20	400	4.472	70	4,900	8.367	120	14,400	10.954
21	441	4.583	71	5,041	8.426	121	14,641	11.000
22	484	4.690	72	5,184	8.485	122	14,884	11.045
23	529	4.796	73	5,329	8.544	123	15,129	11.091
24	576	4.899	74	5,476	8.602	124	15,376	11.136
25	625	5.000	75	5,625	8.660	125	15,625	11.180
26	676	5.099	76	5,776	8.718	126	15,876	11.225
27	729	5.196	77	5,929	8.775	127	16,129	11.269
28	784	5.292	78	6,084	8.832	128	16,384	11.314
29	841	5.385	79	6,241	8.888	129	16,641	11.358
30	900	5.477	80	6,400	8.944	130	16,900	11.402
31	961	5.568	81	6,561	9.000	131	17,161	11.446
32	1,024	5.657	82	6,724	9.055	132	17,424	11.489
33	1,089	5.745	83	6,889	9.110	133	17,689	11.533
34	1,156	5.831	84	7,056	9.165	134	17,956	11.576
35	1,225	5.916	85	7,225	9.220	135	18,225	11.619
36	1,296	6.000	86	7,396	9.274	136	18,496	11.662
37	1,369	6.083	87	7,569	9.327	137	18,769	11.705
38	1,444	6.164	88	7,744	9.381	138	19,044	11.747
39	1,521	6.245	89	7,921	9.434	139	19,321	11.790
40	1,600	6.325	90	8,100	9.487	140	19,600	11.832
41	1,681	6.403	91	8,281	9.539	141	19,881	11.874
42	1,764	6.481	92	8,464	9.592	142	20,164	11.916
43	1,849	6.557	93	8,649	9.644	143	20,449	11.958
44	1,936	6.633	94	8,836	9.695	144	20,736	12.000
45	2,025	6.708	95	9,025	9.747	145	21,025	12.042
46	2,116	6.782	96	9,216	9.798	146	21,316	12.083
47	2,209	6.856	97	9,409	9.849	147	21,609	12.124
48	2,304	6.928	98	9,604	9.899	148	21,904	12.166
49	2,401	7.000	99	9,801	9.950	149	22,201	12.207
50	2,500	7.071	100	10,000	10.000	150	22,500	12.247

Fraction/Decimal Equivalents

Fraction	Decimal	Fraction	Decimal
$\frac{1}{2}$	0.5	$\frac{5}{10}$	0.5
$\frac{1}{3}$	0.3	$\frac{6}{10}$	0.6
$\frac{2}{3}$	0.6	$\frac{7}{10}$	0.7
$\frac{1}{4}$	0.25	$\frac{8}{10}$	0.8
$\frac{2}{4}$	0.5	$\frac{9}{10}$	0.9
$\frac{3}{4}$	0.75	$\frac{1}{16}$	0.0625
$\frac{1}{5}$	0.2	$\frac{2}{16}$	0.125
$\frac{2}{5}$	0.4	$\frac{3}{16}$	0.1875
$\frac{3}{5}$	0.6	$\frac{4}{16}$	0.25
$\frac{4}{5}$	0.8	$\frac{5}{16}$	0.3125
$\frac{1}{8}$	0.125	$\frac{6}{16}$	0.375
$\frac{2}{8}$	0.25	$\frac{7}{16}$	0.4375
$\frac{3}{8}$	0.375	$\frac{8}{16}$	0.5
$\frac{4}{8}$	0.5	$\frac{9}{16}$	0.5625
$\frac{5}{8}$	0.625	$\frac{10}{16}$	0.625
$\frac{6}{8}$	0.75	$\frac{11}{16}$	0.6875
$\frac{7}{8}$	0.875	$\frac{12}{16}$	0.75
$\frac{1}{10}$	0.1	$\frac{13}{16}$	0.8125
$\frac{2}{10}$	0.2	$\frac{14}{16}$	0.875
$\frac{3}{10}$	0.3	$\frac{15}{16}$	0.9375
$\frac{4}{10}$	0.4		